EXAFS
Spectroscopy
Techniques and Applications

EXAFS
Spectroscopy
Techniques and Applications

Edited by

B.K. Teo and D.C. Joy

Bell Laboratories
Murray Hill, New Jersey

PLENUM PRESS · NEW YORK AND LONDON

Library of Congress Cataloging in Publication Data

Main entry under title:

EXAFS spectroscopy, techniques and applications.

   Based on proceedings of a symposium held at the meeting of the Materials Research Society, Nov. 26-30, 1979, in Boston.
   Includes index.
   1. X-ray spectroscopy. 2. Absorption spectra. I. Teo, B. K. II. Joy, David C., 1943-   . III. Materials Research Society.
QC482.S6E9   543'.08586   81-199
ISBN 0-306-40654-3

Based on the proceedings of a symposium on The Applications
of EXAFS to Materials Science, held at the 1979 Meeting of
the Materials Research Society, November 26—30, 1979, in
Boston Massachusetts

© 1981 Plenum Press, New York
A Division of Plenum Publishing Corporation
233 Spring Street, New York, N.Y. 10013

Printed in the United States of America

PREFACE

This book on Extended X-Ray Absorption Fine Structure (EXAFS) Spectroscopy grew out of a symposium, with the same title, organized by us at the 1979 Meeting of the Materials Research Society (MRS) in Boston, MA. That meeting provided not only an overview of the theory, instrumentation and practice of EXAFS Spectroscopy as currently employed with photon beams, but also a forum for a valuable dialogue between those using the conventional approach and those breaking fresh ground by using electron energy loss spectroscopy (EELS) for EXAFS studies.

This book contains contributions from both of these groups and provides the interested reader with a detailed treatment of all aspects of EXAFS spectroscopy, from the theory, through consideration of the instrumentation for both photon and electron beam purposes, to detailed descriptions of the applications and physical limitations of these techniques. While some of the material was originally presented at the MRS meeting all of the chapters have been specially written for this book and contain much that is new and significant.

As will be evident from reading this book, EXAFS measurements using photons as the primary excitation source are well established, even though advances can still be made in the areas of improvement of instrumentations, development of new techniques, understanding of the limitations, and applications to new systems. On the other hand, the use of electrons in EXAFS data acquisition is still under intense development. While it is competitive in terms of counting statistics and potentially very useful for EXAFS of low atomic weight elements and microscopic areas, much needs to be done in ungrading the technique in order to compete or complement the photon techniques.

We hope that this first book on this important topic will be found valuable to those now in the field as well as to those wishing to enter it, including researchers working in the field of physics, chemistry, biology, materials science, and related areas.

Finally, we would like to thank all the authors for their valuable contributions to this book. We are particularly indebted to Professor E. A. Stern and Dr. P. A. Lee, as well as some of the authors, for their generous help in reading some of the chapters.

B. K. Teo and D. C. Joy

Bell Laboratories
Murray Hill, New Jersey

December, 1980

v

CONTENTS

HISTORICAL DEVELOPMENT OF EXAFS

E. A. Stern

University of Washington
Seattle, VA 98195

The Extended X-ray Absorption Fine Structure (EXAFS) has been known for over 50 years, but only recently has its power for structure determination been appreciated. The first experimental detection of fine structure past absorption edges were by Fricke (1920)[1] and Hertz (1920).[2] The first structure detected was the near edge structure (Coster 1924,[3] Lindh 1921,[4] 1925[5]) which could be explained by the theory of Kossel (1920).[6] However, as the experimental measurements extended the detected fine structure to hundreds of eV past the edge (the EXAFS) (Ray 1929,[7] Kievet and Lindsay 1930[8]) a new explanation was required. The temperature dependence in EXAFS was first experimentally noted by Hanawalt (1931).[9,10] Kronig (1931)[11] first attempted an explanation of the EXAFS in condensed matter using the newly developed Quantum Mechanics. His explanation utilized the energy gaps at the Brillouin zone boundaries and thus depended explicitly on the long range order in the solid. Following Azaroff (1963)[12] we will call this theory a long range order (LRO) theory and the other class of theories short range order (SRO). LRO theory is fundamentally in error, but it took over 40 years for the error to be discovered (Stern 1974).[13]

Kronig (1932)[14] also germinated the idea of the SRO theory which he employed to explain EXAFS in molecules. Some elements of the modern theory were missing in his original theory but the basic physical idea was correct. Kronig, apparently, never realized that the same basic physics explains EXAFS in both solids and molecules. Kronig's SRO theory explains EXAFS by the modulations of the final state wave function of the photoelectron caused by the backscattering from the surrounding atoms. Petersen (1932, 1933, 1936)[15-17] developed the Kronig ideas further for molecules by adding the phase shift in the photoelectron wave function caused by both the potentials of the excited atom and the backscattering atoms. The next advance was made by Kostarev (1941, 1949)[18,19] who realized that the Petersen SRO theory was also applicable to matter in the condensed state. The lifetime of the excited photoelectron and core hole state was first accounted for by Sawada et al. (1959)[20] through a mean free path. The remaining missing element was supplied by Shmidt (1961, 1963)[21] who pointed out that the interference of the backscattered waves from atoms at a given average distance (a coordination shell of atoms) will not be all in phase because of the disorder in their distances due to thermal vibrations or structure variations. He introduced a Debye-Waller type factor to account for the thermal disorder based on the Debye theory of lattice vibrations.

The historical development outlined above is not an exhaustive description of all of the contributions to EXAFS theory. The review by Azaroff and Pease (1974)[22] covers the field until 1970 (even though it was published in 1974)[22] and gives a more complete exposition. It is our purpose here to take advantage of hindsight, which was not available at the time of the Azaroff and Pease review, and emphasize just those contributions that we now know are correct. At the time of the above developments, the issue was quite confused. Although the various elements of the modern theory were around, no one put them completely together. In fact, there was even confusion whether the SRO or LRO theory was the correct one. In their review to 1970, nine years after the various elements of the theory had been proposed, Azaroff and Pease (1974)[22] stated: "it is premature to draw any conclusions regarding the most appropriate calculational approach to employ for EXAFS". A clever experimental study of EXAFS which attempted to distinguish between the SRO and LRO theories states in its summary (Perel and Deslattes 1970[23]): "Neither of the (EXAFS) theories discussed here (SRO and LRO) can account for even the 'gross' characteristics of the absorption spectra..."

In spite of this confusion, there were some applications made of EXAFS as an experimental tool. Chemical bond information using near edge structure was obtained by Mitchell and Beaman (1952)[24] and van Nordstrand (1960).[25] Nearest neighbor distance determinations were made by Lytle (1965, 1966)[26,27] and Levy (1965).[28] However, these efforts did not attract general attention because of the confusion surrounding the subject.

A major reason for the confusion was the lack of detailed agreement between any theory and the experiments. The experimental measurements were not always reliable themselves, but this was not a critical factor since there were many reliable measurements available which did not agree with the theories. The SRO theory, though correct in principle, suffered because the atomic parameters that enter it were not calculated accurately.

The situation changed when Sayers, Stern and Lytle (1971)[29] pointed out, based on a theoretical expression of the EXAFS (Sayers et al. 1970)[30] which has since become the accepted modern form, that a Fourier transform of the EXAFS with respect to the photoelectron wave number should peak at distances corresponding to nearest neighbor coordination shells of atoms. The introduction of the Fourier transform changed EXAFS from a confusing scientific curiosity to a quantitative tool for structure determination. Instead of comparing EXAFS measurements to theoretical calculations based on atomic parameters whose values were difficult to calculate, it was now possible to use EXAFS to extract structure information directly and to determine experimentally all of the required atomic parameters. The correctness of the SRO theory was now obvious since the transform revealed only the first few nearest neighbor shells of atoms.

The accessibility of EXAFS measurements was greatly enhanced with the availability of synchrotron radiation sources of X-rays several years after the potential of EXAFS was first shown. Because synchrotron sources typically have x-ray intensities 3 or more orders of magnitude greater in the continuum energies than do the standard x-ray tube sources, the time for measuring a spectrum for concentrated samples dropped from the order of a week (Lytle et al. 1975)[31] to the order of minutes. At the same time, these sources expanded possibilities by making feasible the measurement of dilute samples which could not be even contemplated before.

REFERENCES

1. H. Fricke, Phys. Rev. **16,** 202 (1920).

2. G. Hertz, Zeit. f. Physik, **3,** 19 (1920).

3. D. Coster, Zeit. f. Physik, **25,** 83 (1924).

4. A. E. Lindh, Zeit. f. Physik **6,** 303 (1921).

5. A. E. Lindh, Zeit. f. Physik **31,** 210 (1925).

6. W. Kossel, Zeit. f. Physik **1,** 119; **2,** 470 (1920).

7. B. B. Ray, Zeit. f. Physik **55,** 119 (1929).

8. B. Kievit and G. A. Lindsay, Phys. Rev. **36,** 648 (1930).

9. J. D. Hanawalt, Zeit. f. Physik **70,** 20 (1931).

10. J. D. Hanawalt, Phys. Rev. **37,** 715 (1931).

11. R. de L. Kronig, Z. Physik **70,** 317 (1931).

12. L. V. Azaroff, Rev. Mod. Phys. **35,** 1012 (1963).

13. E. A. Stern, Phys. Rev. **B10,** 3027 (1974).

14. R. de L. Kronig, Zeit. f. Physik **75,** 468 (1932).

15. H. Peterson, Zeit. f. Physik **76,** 768 (1932).

16. H. Peterson, Zeit. f. Physik **80,** 528 (1933).

17. H. Peterson, Zeit. f. Physik **98,** 569 (1936).

18. A. I. Kostarev, Zh. Eksper. Teor. Fiz. **11,** 60 (1941).

19. A. I. Kostarev, Zh. Eksper. Teor. Fiz. **19,** 413 (1949).

20. M. Sawada, Rep. Sci. Works Osaka Univ. **7,** 1 (1959).

21. V. V. Shmidt, Bull. Acad. Sci. USSR, Ser. Phys. **25,** 998 (1961); V. V. Shmidt, ibid., **27,** 392 (1963).

22. L. V. Azaroff and D. M. Pease, X-ray Absorption, in: "X-ray Spectroscopy," L. V. Azaroff, ed., McGraw-Hill, N.Y. ch. 6 (1974).

23. J. Perel and R. D. Deslattes, Phys. Rev. **B2,** 1317, (1970).

24. G. Mitchell and W. W. Beeman, J. Chem. Phys. **20,** 1298 (1952).

25. R. A. Van Nordstrand, Advances in Catalysis **12,** 149 (1960).

26. F. W. Lytle in: "Physics of Non-Crystalline Solids", J. A. Prins, ed., North-Holland, Amsterdam, p. 12 (1965).

27. F. W. Lytle, Adv. X-ray Analysis **9,** 398 (1966).

28. R. M. Levy, J. Chem. Phys. **43,** 1846 (1965).

29. D. E. Sayers, E. A. Stern and F. W. Lytle, Phys. Rev. Lett. **27,** 1204 (1971).

30. D. E. Sayers, F. W. Lytle and E. A. Stern, Adv. X-ray Anal. **13,** 248 (1970).

31. F. W. Lytle, D. E. Sayers and E. A. Stern, Phys. Rev. **B11,** 4825 (1975).

THEORY OF EXTENDED X-RAY ABSORPTION FINE STRUCTURE

P. A. Lee

Bell Laboratories
Murray Hill New Jersey   07974

Extended X-ray Absorption Fine Structure (EXAFS) refers to oscillations of the X-ray absorption coefficient on the high energy side of an absorption edge. Such oscillations can extend up to 1000 eV above the edge and may have a magnitude of 10% or more. This phenomenon has been known for half a century and the basic physical explanation has been provided by Kronig[1] as being due to modification of the final state of the photoelectron by atoms surrounding the excited atom. More recent work has established that a single scattering short-range order theory is adequate under most circumstances. The oscillatory part of the absorption coefficient $\Delta\chi$ normalized to the background χ_0 is given by [2-4]

$$\frac{\Delta\chi}{\chi_0} = - \sum_i 3(\hat{\epsilon}.\hat{r}_i)^2 \, F(k) \, \frac{\sin(2kr_i + 2\delta'_1(k) + \theta(k))}{kr_i^2} e^{-2\sigma_i^2 k^2} e^{-2r_i/\lambda} \tag{1}$$

for excitations of an s state in a system by X-ray polarized in the $\hat{\epsilon}$ direction. Eq. 1 describes the modification of the electron wave function at the origin due to scattering by N_i neighbors located at a radial distance r_i away. The scattering amplitude is given by

$$\begin{aligned} f(\pi) &\equiv F(k)e^{i\theta(k)} \\ &= \frac{1}{2ik} \sum_l (2l+1)(e^{2i\delta_l}-1)(-1)^l \end{aligned} \tag{2}$$

where δ_l are the scattering phase shifts. The electron wave vector k is defined as

$$k = (2m/\hbar^2(h\nu-E_0))^{\frac{1}{2}} \tag{3}$$

where ν is the X-ray frequency and E_0 is some choice of the threshold energy. It is clear that the electron wave will be phase shifted by $2kr_i$ by the time it makes the return trip to the neighbor. To this we must add the phase change in the backscattering process $\theta(k)$ and also twice the central atom phase shift δ'_1 (the prime denotes the fact that the central atom is photoexcited and is in general different from the neighbor) to account for the potential of the central atom that the $l=1$ photoelectron wave has

5

traversed. The wave function at the origin is therefore modulated according to this total phase factor and this accounts for the sinusoidal term in Equation 1. In addition we have to account for the fact that the atoms are not stationary. Thermal vibration will smear out the EXAFS oscillation and in the harmonic approximation this can be accounted for by a Debye-Waller type term $\exp(-2\sigma_i^2 k^2)$. Finally electrons that have suffered inelastic losses will not have the right wave vector to contribute to the interference process. This is crudely accounted for by an exponential damping term $exp(-2r_i/\lambda)$ where λ is the inelastic electron mean free path. It is the limited range of the photoelectrons in the energy region of interest (50-1,000 eV) that permits a short-range order description of EXAFS even in crystalline materials.

A comprehensive review of EXAFS with extensive discussion of the theory is under preparation[5], so we will only summarize here the key features of the theory that has been developed in the last few years. The two aspects of the theory are (i) to determine whether the simple formula given by Eq. 1 is adequate and (ii) to perform ab initio calculation of the functions $F(k)$, $\delta'_1(k)$ and $\theta(k)$ for atoms throughout the periodic table and to test them against experiments. Part (ii) of this program has been carried out by Teo and Lee[6] based on earlier theoretical development by Lee and Beni.[7] Some of their results are shown in Figures 1-3. Overall it is found that if the threshold energy E_o is allowed to vary as an adjustable parameter, the calculated phase shifts are in good agreement with experiments so that nearest neighbor distances can be determined with an accuracy of about 0.02Å. The calculated amplitude $F(k)$ is found to be too large by as much as a factor of two in some cases, and appears to depend on the atomic number Z and its chemical environment. We shall return to discuss this later.

Going back to the question of the adequacy of Eq. 1, the following issues have been addressed:

a. Multiple scattering:

Equation 1 takes into account only a single backscattering. It is clear EXAFS is an interference phenomenon which bears obvious relations to LEED and angular resolved photoemission. Indeed EXAFS can be thought of as LEED with a source and a detector located at the central atom. Since multiple scattering is known to be very important in LEED, the natural question is why should the single scattering formula, Eq. 1, be at all adequate. The answer is that one can account for multiple scattering events by adding all scattering paths that originate and terminate at the central atom.[3] This is illustrated in Fig. 4. Each of these events behaves like $sin(2kr_{eff})$ where $2r_{eff} = r_1+r_2+r_3$ is the total path length. Clearly r_{eff} is large compared with the nearest and perhaps the next nearest neighbor spacing. Thus multiple scattering events give rise to rapidly oscillatory contribution in k space which tend to cancel out.

b. Chemical transferability of phases:

Equation 1 separates the scattering problem into scattering by individual atoms. All effects of chemical bonding, nonspherical potential, atomic configuration and ionicity are ignored. It has been shown[7] that all such effects have a decreasing influence on the phase shift functions for increasing energy, and can be mimicked by a change in the threshold energy E_o. The observation is the basis of the success of the calculated phase shift in the prediction of bond lengths.

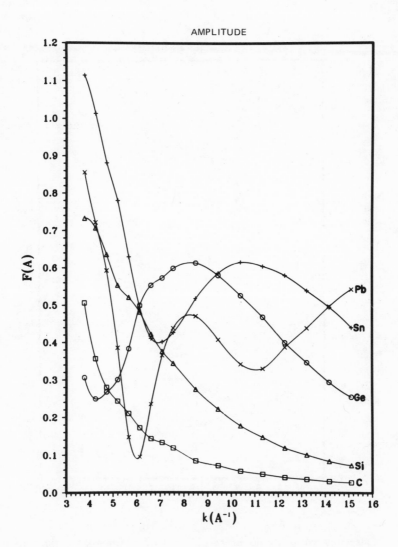

Fig. 1   Calculated backscattering amplitude $F(k)$ for a number of elements (from Ref. 6).

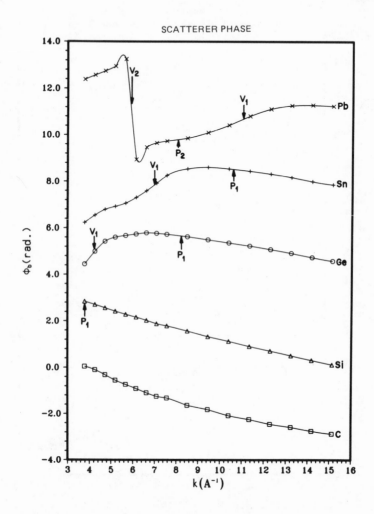

Fig. 2   Calculated backscattering phase $\phi_b = \theta(k)$ for a number of elements (from Ref. 6).

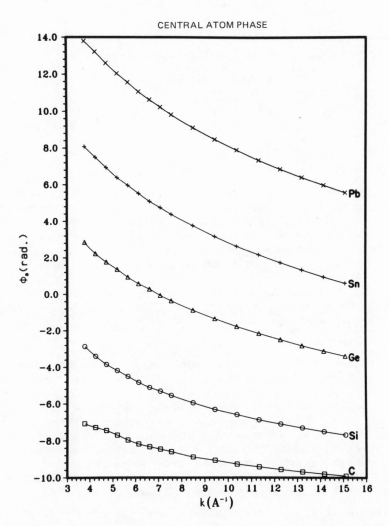

CENTRAL ATOM PHASE

Fig. 3   Calculated central atom phase shift $\phi_a = 2\delta'_1(k)$ for a number of elements (from Ref. 6).

Fig. 4.   Illustration of a multiple scattering event.

c. Thermal and static disorder:

Equation 1 assumes that disorder can be accounted for by a Debye-Waller type factor $exp(-2\sigma^2 k^2)$. This assumes that the disorder is Gaussian or that the potential is harmonic in the case of thermal disorder. Important exceptions to this have been noticed in the case of systems with large thermal disorder, such as Zn at high temperature[8] or in systems with large intrinsic disorder, such as superionic conductors[9] or liquids. It is found that the effects of disorder may show up as an overall reduction in the EXAFS amplitude rather than as a Debye-Waller factor. Such effects could lead to serious difficulties in attempts to measure the number of nearest neighbors. At the same time, deviation from Gaussian behavior produces important corrections to the phase shift.[8]

d. Amplitude transferability:

In Equation 1 inelastic losses are modelled by an exponential damping term $exp(-2r_i/\lambda)$. This is certainly inadequate. We have already mentioned the discrepancy between the calculated amplitude $F(k)$ and the experiment. A source of this discrepancy has been determined to be core hole relaxation processes[7,10] in the central atom. In addition we have inelastic losses by the photoelectron in the backscattering process and as it traverses the material. The loss mechanism is generally the excitation of low lying states in the system, such as plasmons or interband transitions which will be sensitive to the chemical environment. Thus the amplitude is not expected to have the same degree of chemical transferability as the phase, a conclusion reached by experimental studies as well.[11,12]

In conclusion the basic physics leading to a simple theoretical description of EXAFS is understood and tested against experiments. Conditions under which the simple theory as given by Eq. 1 fails has also been investigated. These results indicate that care must be taken in the interpretation of EXAFS in highly disordered systems or in attempts to determine co-ordination number.

REFERENCES

1. R. de L. Kronig, Z. Phys. **70**, 317 (1931); **75**, 468 (1932).

2. D. E. Sayers, E. A. Stern, and F. W. Lytle, Phys. Rev. Lett. **27**, 1207; (1971) Phys. Rev. **B11**, 4836 (1975) and references therein.

3. P. A. Lee and J. B. Pendry, Phys. Rev. **B11**, 2795 (1975).

4. C. Ashley and S. Doniach, Phys. Rev. **B11**, 1279 (1975).

5. P. A. Lee, P. H. Citrin, P. Eisenberger, and B. M. Kincaid, to be submitted to Rev. Mod. Phys.

6. B. K. Teo and P. A. Lee, J. Am. Chem. Soc. **101**, 2815 (1979).

7. P. A. Lee and G. Beni, Phys. Rev. **B15**, 2862 (1977).

8. P. Eisenberger and G. Brown, Solid State Comm. **29**, 481 (1979).

9. T. M. Hayes, J. B. Boyce, and J. L. Beeby, J. Phys. **C11**, 2931 (1978).

10. J. J. Rehr, E. A. Stern, P. L. Martin, and E. R. Davidson, Phys. Rev. **B17**, 560 (1978).

11. P. Eisenberger and B. Lengeler, preprint.

12. E. A. Stern, B. A. Bunker, and S. M. Heald, Phys. Rev. **B21,** 5521 (1980).

EXTENDED X-RAY ABSORPTION FINE STRUCTURE (EXAFS)

SPECTROSCOPY: TECHNIQUES AND APPLICATIONS

Boon-Keng Teo

Bell Laboratories
Murray Hill, New Jersey 07974

Extended X-ray absorption fine structure (EXAFS) refers to the oscillatory variation of the X-ray absorption as a function of photon energy beyond an absorption edge. The absorption, normally expressed in terms of absorption coefficient (μ), can be determined from a measurement of the attenuation of X-rays upon their passage through a material. When the X-ray photon energy (E) is tuned to the binding energy of some core level of an atom in the material, an abrupt increase in the absorption coefficient, known as the absorption edge, occurs. For isolated atoms, the absorption coefficient decreases monotonically as a function of energy beyond the edge. For atoms either in a molecule or embedded in a condensed phase, the variation of absorption coefficient at energies above the absorption edge displays a complex fine structure called EXAFS.

Although the extended fine structure has been known for a long time,[1] its structural content was not fully recognized until the recent work of Stern, Lytle, and Sayers.[2] In addition, the recent availability of synchrotron radiation has resulted in the establishment of EXAFS as a practical structural tool.[3] This technique is especially valuable for structural analyses of chemical or biological systems where conventional diffraction methods are not applicable.

In this chapter, a brief description of this powerful structural method and the underlying physics will be given, followed by illustrations of its application to chemical and biochemical systems. The physics and materials aspects of EXAFS can be found elsewhere in this book. Other comprehensive reviews can also be found in the literature.[4]

Introduction

From a qualitative viewpoint, the probability that an X-ray photon will be absorbed by a core electron depends on both the initial and the final states of the electron. The initial state is the localized core level corresponding to the absorption edge. The final state is that of the ejected photoelectron which can be represented as an outgoing spherical wave originating from the X-ray absorbing atom. If the absorbing atom is surrounded by a neighboring atom, the outgoing photoelectron wave will be backscattered by the neighboring atom, thereby producing an incoming electron wave. The final state

13

is then the sum of the outgoing and all the incoming waves, one per each neighboring atom. It is the interference between the outgoing and the incoming waves that gives rise to the sinusoidal variation of μ vs E known as EXAFS (*cf.* Figure 1).

The frequency of each EXAFS wave depends on the distance between the absorbing atom and the neighboring atom since the photoelectron wave must travel from the absorber to the scatterer and back. During the trip, the photoelectron actually experiences a phase shift (Coulombic interaction) of the absorber twice (i.e., once going out and once coming back) and a phase shift of the scatterer once (scattering). If one assumes that the phase shifts can be obtained from either model compounds or calculations, one can determine interatomic distances in the vicinity of the absorber. On the other hand, the amplitude of each EXAFS wave depends upon the number and the backscattering power of the neighboring atom, as well as on its bonding to and distance from the absorber (*vide infra*). From an analysis of the scattering profiles, one can quantitatively assess the types and numbers of atoms surrounding the absorber.

Structural determinations via EXAFS depend on the feasibility of resolving the data into individual waves corresponding to the different types of neighbors of the absorbing atom. This can be accomplished by either curve-fitting or Fourier transform techniques. Curve-fitting involves a best fitting of the data with a sum of individual waves modeled by some empirical equations, each of which contains appropriate structural parameters for each type of neighbor. On the other hand, the Fourier transform technique provides a photoelectron scattering profile as a function of the radial distance from the absorber. In such a radial distribution function, the positions of the peaks are related to the distance between the absorber and the neighboring atoms while the sizes of the peaks are related to the numbers and types of the neighboring atoms.

To appreciate the usefulness of the EXAFS technique, we need only compare it with diffraction methods. For crystalline materials with long-range ordering, their structures are generally determined by X-ray or neutron diffraction in which measurements of the diffracted intensities (Bragg's law) can yield a three-dimensional picture of atomic coordinates via a Fourier transformation. For materials with only short-range order (amorphous solid, liquid, or solution) X-ray scattering experiments provide only diffuse halos which, upon Fourier transformation, give rise to a one-dimensional (1-D) radial distribution function (RDF). Such a RDF contains interatomic distances due to all atomic pairs in the sample (condensed to one arbitrary origin). On the other hand, EXAFS measurements can provide structural information for each type of atom if one simply tunes the X-ray energy to coincide successively with an absorption edge of each of the atom types in the sample. Such information as the number and kind of neighboring atoms and their distances away from the absorber are contained in the 1-D RDF centered at the absorber. It is clear that EXAFS is highly specific in that it can focus on the immediate environment around each absorbing species (generally out to *ca* 6 Å corresponding to 1-3 coordination shells). Other materials or impurities present in the sample which either do not contain the absorber or are not directly bound to the absorber will not interfere. Furthermore, the technique is highly versatile in that it can be applied with about the same degree of accuracy (0.01 ~0.03Å) to matter in the solid (crystalline or amorphous), liquid, solution, or gaseous state. The following sections attempt to give a more formal but highly simplified description of the method and its application.

EXAFS Spectroscopy

EXAFS spectroscopy refers to the measurement of the X-ray absorption coefficient μ as a function of photon energy E above the threshold of an absorption edge. Figure 1 shows schematically one edge of an absorber. In a transmission experiment, μ or μx (x is the sample thickness) is calculated by

$$\mu x = \ln I_o/I \tag{1}$$

where I_o and I are the intensities of the incident and transmitted beams, respectively.

EXAFS spectra generally refer to the region 40-1000 eV above the absorption edge. Near or below the edge, there generally appear absorption peaks due to excitation of core electrons to some bound states ($1s$ to nd, $(n + 1)s$, or $(n + 1)p$ orbitals for K edge, and $2s$ for L_I edge, $2p$ for L_{II}, L_{III} edges to the same set of vacant orbitals, etc.). This pre-edge region contains valuable bonding information such as the energetics of virtual orbitals, the electronic configuration, and the site symmetry.[5,6] The edge position also contains information about the charge on the absorber.[5,6] The region in between is least understood. It arises from effects such as many-body interactions, multiple scatterings, distortion of the excited state wavefunction by the Coulomb field, band structures, etc.

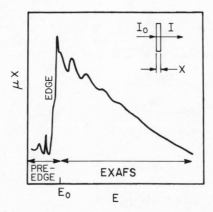

Fig. 1.   Schematic representation of the transmission experiment and the resulting x-ray absorption spectrum μx vs E for an absorption edge of an absorber in a molecule (e.g. K edge of iron which corresponds to the ejection of a $1s$ electron by absorption of an X-ray photon with $E \geq E_0$, where E_0 is the energy threshold).

Transmission is just one of several modes of EXAFS measurements. The fluorescence technique involves the measurement of the fluorescence radiation (over some solid angle) at right angle to the incident beam.[7a−c] For dilute biological systems, this method removes the 'background' absorption due to other constituents, thereby improving the sensitivity by orders of magnitude. Both of these methods require no vacuum technique. Other more specialized methods include: (1) surface EXAFS (SEXAFS) studies which involve measurements (in vacuum) of either the Auger electrons[7d] or a portion of the inelastically scattered electrons (partial electron yield)[7e] during the relaxation of an atom that has absorbed a photon; and (2) electron energy loss (inelastic electron scattering) spectroscopy (EELS).[7f−j] These later methods are useful for light atom EXAFS with edge energies up to a few keV. These experimental techniques will be described in a later section.

It should be emphasized that in the transmission or fluorescence mode, EXAFS spectroscopy involves only X-ray measurements despite the fact that concepts such as the photoelectric effect and the scattering of electrons by atoms are invoked in the theory of EXAFS. For the other experimental modes, on the other hand, electrons may be involved in the experimentation as we shall see in a later section.

Theory

EXAFS is a final state interference effect involving scattering of the outgoing photoelectron from the neighboring atoms. Figure 2 attempts to convey, pictorially, the

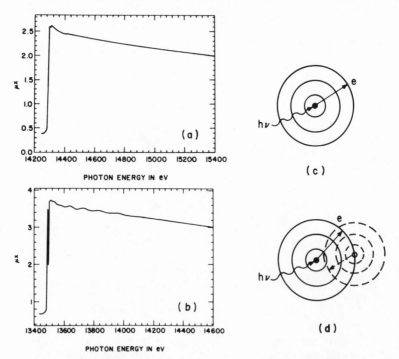

Fig. 2.   Qualitative rationalization of the absence and presence, respectively, of EXAFS in a monatomic gas such as *Kr* (a and c) and a diatomic gas such as *Br*$_2$ (b and d).

current view of EXAFS. For a monoatomic gas such as Kr (Fig. 2a and c)[3a] with no neighboring atoms, the ejection of a photoelectron by absorption of a X-ray photon will travel as a spherical wave with a wavelength $\lambda = \dfrac{2\pi}{k}$ where

$$k = \sqrt{\frac{2m}{\hbar^2}(E - E_o)} \tag{2}$$

Here E is the incident photon energy and E_o is the threshold energy of that particular absorption edge. The μ vs E curve follows the usual smooth λ^3 decay (cf. Fig. 2a). In the presence of neighboring atoms (e.g., in Br_2, Fig. 2b and d),[3a] this outgoing photoelectron can be backscattered from the neighboring atoms thereby producing an incoming wave which can interfere either constructively or destructively with the outgoing wave near the origin, resulting in the oscillatory behavior of the absorption rate (cf. Fig. 2b). The amplitude and frequency of this sinusoidal modulation of μ vs E depend on the type (and bonding) of the neighboring atoms and their distances away from the absorber, respectively.

This simple picture of EXAFS has been formulated into the generally accepted *short-range single-electron single-scattering* theory.[2,8] For reasonably high energy ($\gtrsim 60$ eV) and moderate thermal or static disorders, the modulation of the absorption rate in EXAFS, normalized to the 'background' absorptions (μ_o)[9] is given by

$$\chi(E) = \frac{\mu(E) - \mu_o(E)}{\mu_o(E)} \tag{3}$$

In order to relate $\chi(E)$ to structural parameters, it is necessary to convert the energy E into the photoelectron wavevector k via Eq. 2. This transformation of $\chi(E)$ in E space gives rise to $\chi(k)$ in k space where

$$\chi(k) = \sum_j N_j S_i(k) F_j(k) e^{-2\sigma_j^2 k^2} e^{-2r_j/\lambda(k)} \frac{Sin(2kr_j + \phi_j(k))}{kr_j^2} \tag{4}$$

Here $F_j(k)$ is the backscattering amplitude from each of the N_j neighboring atoms of the jth type with a Debye-Waller factor of σ_j (to account for thermal vibration (assuming harmonic vibration) and static disorder (assuming Gaussian pair distribution) and at a distance r_j away. $\phi_j(k)$ is the total phase shift experienced by the photoelectron. The term $e^{-2r_j/\lambda}$ is due to inelastic loses in the scattering process (due to neighboring atoms and the medium in between) with λ being the electron mean free path. $S_i(k)$ is the amplitude reduction factor due to many-body effects such as shake up/off processes at the central atom. It is clear that each EXAFS wave is determined by the backscattering amplitude $(N_j F_j(k))$, modified by the reduction factors $S_i(k)$, $e^{-2\sigma_j^2 k^2}$, and $e^{-2r_j/\lambda}$, and the $1/kr_j^2$ distance dependence, and the sinusoidal oscillation which is a function of interatomic distances $(2kr_j)$ and the phase shift $(\phi_j(k))$.

It should be emphasized that while the amplitude function $F_b(k)$ depends only on the type of the backscatters (except, perhaps, the reduction factor $S(k)$ which is mainly a function of the absorber) the phase function contains contributions from both the absorber and the backscatterer.

$$\phi_{ab}^\ell(k) = \phi_a^\ell(k) + \phi_b(k) - \ell\pi \tag{5}$$

where $\ell = 1$ for K and L_I edges and $\ell = 2$ or 0 for $L_{II,III}$ edges. Here $\phi_a = 2\delta_\ell'$ is the ℓ phase shift of the absorber and $\phi_b = \theta$ is the phase of the backscattering amplitude.[8]

Qualitatively, the physical origin of this dependence is that the photoelectron experiences the central atom phase shift twice, once going out and once coming back, but experiences the neighboring atom phase shift once by propagating from the absorber to the neighboring atoms and back to the absorber. The subscripts a and b in Eq. 5 refer to the subscripts i and j of Eq. 4.

Thermal and Static Disorder Effects

The Debye-Waller factor σ plays an important role in EXAFS spectroscopy.[10] It contains important structural and chemical information which is otherwise difficult to obtain, yet it comes as a bonus in the EXAFS determination of interatomic distance. Generally speaking, the Debye-Waller factor σ has two components σ_{stat} and σ_{vib} due to static disorder and thermal vibrations, respectively. Assuming small disorders with a symmetric pair distribution function for static disorder and harmonic vibration for thermal disorder,

$$\sigma^2 = \sigma_{stat}^2 + \sigma_{vib}^2 \tag{6}$$

In principle, these two factors can only be separated by a temperature dependent study of $\sigma(T)$. However, if σ_{vib} can be estimated from vibrational spectroscopies or if σ_{stat} is known from other studies, the other term can be calculated from the experimentally determined σ.

The term σ_{stat} is related to the symmetric Gaussian pair distribution function of distance r

$$g(r) = \frac{1}{\sqrt{2\pi}\,\sigma} e^{-(r-r_o)^2/2\sigma^2} \tag{7}$$

where σ stands for σ_{stat} and $r - r_o$ is the deviation from the mean distance r_o. For discrete bonds, σ_{stat} is related to the *root-mean-square* standard deviation δ (keeping terms to second order):

$$\sigma_{stat} \approx \delta = \sqrt{\sum_{j=1}^{N} \frac{(r_j^2 - r_0)^2}{N}} \tag{8}$$

For a two-distance system with m bonds at a distance r_m and n bonds at a distance r_n it reduces to

$$\sigma_{stat} \approx \frac{\sqrt{mn}}{m+n} \Delta r = \frac{\sqrt{mn}}{m+n} |r_m - r_n| \tag{9}$$

For systems with large difference in discrete distances, particularly those with interference patterns (beat nodes), it is possible to separate distances from Debye-Waller factors as well as to determine the signs of the differences of both *(vide infra)*.

Assuming harmonic motion for a diatomic system, the vibrational contribution to the Debye Waller factor is given by

$$\sigma_{vib}^2 = \frac{h}{8\pi^2\mu\nu} Coth \frac{h\nu}{2kT} \tag{10}$$

where μ is the reduced mass, T is the temperature, and ν is the vibrational frequency.[11] It can be used as a qualitative assessment of bond strength for closely related systems assuming that ν is a qualitative measure of bond strength. For μ in atomic mass units and $\bar{\nu} = \nu/c$ in cm^{-1}, and T in °K, σ_{vib} in Å is given by

$$\sigma_{vib} = 4.106 \left[\frac{1}{\mu \bar{\nu}} \; Coth \; \left[\frac{x}{2} \right] \right]^{1/2} \tag{11a}$$

where $x = 1.441 \dfrac{\bar{\nu}}{T}$. For strong bonds or at the low temperature limit $h\nu >> kT$, $x >> 1$ (typically $x \gtrsim 2$)

$$\sigma \approx 4.106 \left[\frac{1}{\mu \bar{\nu}} \right]^{1/2} = 3.151 \times 10^{-3} \left[\frac{\bar{\nu}}{K} \right]^{\frac{1}{2}} \tag{11b}$$

For weak bonds or at the high temperature limit $h\nu << kT$, $x << 1$ (typically $x \lesssim 0.3$):

$$\sigma \approx 4.836 \left[\frac{T}{\mu \bar{\nu}^2} \right]^{1/2} = 3.712 \times 10^{-3} \left[\frac{T}{K} \right]^{\frac{1}{2}} \tag{11c}$$

Here $K = 4\pi^2 \mu c^2 \bar{\nu}^2$ in mdyne/Å is the force constant. For weak bonds, it is often necessary to do the experiments at low temperatures so as to reduce σ_{vib}, thereby enhancing the EXAFS signal, especially at high k values. For detailed discussions on the determination and calculation of EXAFS σ, see papers by G. Beni and P. M. Platzman;[12] R. B. Greegor and F. W. Lytle;[13] E. Sevillano, H. Meuth, and J. J. Rehr.[14]

For systems with large disorders ($\sigma \gtrsim 0.1$ Å), the above treatments are inadequate.[15–19] For large static disorder, the EXAFS expression must be averaged over a pair distribution function $g(r)$ characteristic of the system (which may well be an asymmetric distribution of interatomic distances):

$$\chi(k) = \frac{F(k)}{k} \int_0^{\infty} g(r) e^{-2r/\lambda(k)} \; \frac{Sin(2kr + \phi(k))}{r^2} \; dr \tag{12}$$

Generally speaking, a large disorder effect is often accompanied by either an asymmetric pair correlation function (such as in the case of amorphous or liquid metals) or an anharmonic vibration potential (such as in the case of zinc). The former is a static disorder which is an intrinsic property whereas the latter is a thermal disorder which can be reduced by lowering the temperature. Failure to take into account these two types of large disorder effects will lead to serious errors. Large disorder can lead to a reduction of the EXAFS amplitude, and hence the apparent coordination numbers, up to an order of magnitude. It can also cause an apparent contraction (or less frequently, expansion) in the near neighbor distances which can be as large as 0.15 Å.[15]

The physical reason is that the broad tail(s) of a pair distribution function will contribute to EXAFS only at low k values due to the exponential damping effect. A large part of this information is lost owing to a large $k_{min} \gtrsim 3$ Å^{-1} cut-off of the data necessitated by the edge complication. The truncated EXAFS therefore only contains structural information characteristic of the sharpest feature/edge in $g(r)$. In most cases, the asymmetry in $g(r)$ arises from hard-core repulsion which has a sharp rise in $g(r)$ at a low distance and a long tail at larger distances. The sharp rise in $g(r)$ gives rise to the dominant feature in EXAFS which persists to high k. In contrast, the broad tail in $g(r)$ contributes to EXAFS only at low k values, much of which is lost due to truncation at $k_{min} \approx 3$ Å^{-1}. It is then not surprising to find an apparent reduction in amplitude and an apparent shrinkage in the distance.[15] For these systems, $g(r)$ must be properly modelled. The EXAFS data, in either k or r space, can then be fitted with an appropriate expression based on the model. The results, with appropriate amplitude and phase corrections, are to some extent model dependent. The modelling of $g(r)$ for a number of different systems will be discussed by Hayes and Boyce[16,17] as well as by Crozier elsewhere in this book.[18,19]

For large thermal disorder, the EXAFS must be integrated over the Boltzman distribution

$$g(r) = Ce^{-U(r)/k_B T} \tag{13}$$

where $U(r)$ is the interatomic potential (which may well be anharmonic) and C is a normalization constant. Even with a harmonic potential $U(r)$, it was shown by Eisenberger and Brown[15] that there should still be a phase shift correction of $\tan^{-1}[-2\sigma^2(k)(1/r_0 + 1/\lambda(k))k]$ if r deviates significantly from r_0, the average distance (note that the factorization of $e^{-2\sigma^2 k^2}$ and $e^{-2r/\lambda(k)}$ in the EXAFS expression is only possible if one assumes $r \approx r_0$ such that the factor $e^{-2r_0/\lambda(k)}/r_0^2$ can be taken out of the integral). Normally, however, this amounts to only $\lesssim 0.01$ Å in distance correction. With an anharmonic potential $U(r)$, serious phase and amplitude corrections can occur. A case in point is the room temperature EXAFS of the c-direction of single crystal Zn where a conventional analysis using the low temperature (20°K) data as a model predicts an "apparent" *contraction* of 0.09 Å of the nearest $Zn-Zn$ distance at room temperature, contrary to the known *expansion* of 0.05 Å.[15]

A generalized phenomenological expression of EXAFS is then

$$\chi(k) = \sum_j N_j S_i(k) F_j(k) \sqrt{A_j^2 + S_j^2} \, e^{-2r_j/\lambda(k)} \frac{Sin\,(2kr_j + \phi_j(k) + Tan^{-1}(A_j/S_j))}{kr_j^2} \tag{14}$$

where

$$S_j(k,r_j,T) = \int_{-\infty}^{\infty} \frac{g_j(x,r_j,T)\,Cos\,2kx}{(1+x/r_j)^2}\,dx \tag{15}$$

$$A_j(k,r_j,T) = \int_{-\infty}^{\infty} \frac{g_j(x,r_j,T)\,Sin\,2kx}{(1+x/r_j)^2}\,dx \tag{16}$$

Here $x = r - r_j$ is the deviation of the distance r from the mean distance r_j. S_j and A_j are potential functions of k, r_j (static disorder), and T (thermal disorder).[15] For a symmetric Gaussian pair correlation function and/or a harmonic thermal vibration and assuming $x \ll r_j$, $A_j = 0$ and $S_j = e^{-2\sigma^2 k^2}$ where σ is given by Eq. 7 and 10, respectively. Equation 14 then reduces to Equation 4.

If nothing is known about the pair distribution function, as pointed out by Eisenberger and Brown,[15] one can simply expand sin $2kx$ and cos $2kx$ and keep the first order terms to derive expressions for $A_j(k)$ and $S_j(k)$ in terms of the moments of $g_j(r)$.

Inelastic Scatterings

Broadly speaking, there are two categories of inelastic scattering processes which tend to reduce the EXAFS amplitude. The first is caused by multiple excitations at the central atom whereas the second is associated with excitation of the neighboring environment, including the neighboring atoms and the intervening medium, by the photoelectron.

In Equations 4 and 14, the inelastic losses due to multiple excitations at the absorber is approximated by an amplitude reduction factor $S_i(k) \lesssim 1$. A better approximation is to use instead the overlap of the initial and final state wave functions of the $(N - 1)$ electrons, the so-called passive electrons which are excited along with the photoelectron:

$$s_0^2(k) = |< \phi_{N-1}'|\phi_{N-1} >|^2 < 1 \qquad (17)$$

where ϕ_{N-1} and ϕ_{N-1}' are the wave functions of the $N-1$ electrons before and after the photoexcitation, respectively. The physical origin of this loss mechanism is that the excess energy $(E - E_0)$ in the photoionization process can excite (shake-up) or ionize (shake-off) other low-binding (outer) electrons within the central atom. In general, these multiple excitations have a broad spectrum which tend to wash out the EXAFS signal. In most cases, $s_0^2(k) \approx 1$ at low k values and $s_0^2(k) \approx 0.6 - 0.8$ for $k \gtrsim 7\text{Å}^{-1}$. It is interesting to note that $s_0^2(k)$ is the only term in the EXAFS amplitude which is significantly dependent upon the central atom. The effect of $s_0^2(k)$ on the amplitude envelope[8b,20,21] will be discussed in detail in a later chapter by Stern, Bunker, and Heald.[21]

In Eq. 4 and 14, the inelastic losses due to excitation of the neighboring environment is approximated by $e^{-2r/\lambda(k)}$ where $\lambda(k)$, which depends on k, is the electron inelastic mean free path. Strictly speaking, one should instead use the expression $L_i(k)L_j(k)L_{ij}(r,k)$ where $L_i(k)$, $L_j(k)$, and $L_{ij}(r,k)$ are inelastic losses due to excitations of the medium, $L_{ij}(r,k)$, as the electron travels from the central atom, $L_i(k)$, to the neighboring atom j, $L_j(k)$, and back.[22] Clearly, the exponential damping term $e^{-2r_j/\lambda(k)}$ approximates, albeit insufficiently, only $L_{ij}(r,k)$ for higher shells. Furthermore, since inelastic scattering processes are dominated by weakly bound electrons (outer electrons such as conduction and valence electrons) which are highly dependent upon chemical environment, we expect these loss mechanisms to be somewhat chemical sensitive, especially for low Z (atomic number) scatterers where the number of valence electrons is a substantial portion of the total number of electrons (e.g. $Z \lesssim 10$) and for higher shells where the emitted photoelectron must travel a greater distance.

Granting the exponential damping form $e^{-2r/\lambda(k)}$, the electron inelastic mean free path can be approximated by[8b,23]

$$\lambda(k) = k^n/\eta \qquad (18)$$

where $n = 1$ or 2 (our experience showed that the best fits occur for $1 < n < 2$ in most cases) for EXAFS data with $5 \lesssim k \lesssim 13 \text{ Å}^{-1}$. If EXAFS data with $k < 5 \text{ Å}^{-1}$ are included, a better model[23] will be

$$\lambda(k) = \frac{1}{\eta}\left[k^n + \left[\frac{\xi}{k}\right]^4\right] \qquad (19)$$

(where η and ξ are constants obtainable by fitting the EXAFS data) since it has been shown by Powell,[24] Penn,[25] Seah and Dench[26] that $\lambda(E) \propto E$ at high energies $(E \gtrsim 500\text{eV})$, $\propto E^{1/2}$ at intermediate energies $(100 \lesssim E \lesssim 500\text{eV})$, and $\propto E^{-2}$ at low energies $(E \lesssim 50\text{eV})$. We find, however, for EXAFS data with $k \gtrsim 5 \text{ Å}^{-1}$, Eq. 18 is adequate in most cases.

Multiple Scatterings

The single-electron single-scattering theory of EXAFS discussed thus far makes use of the fact that in most cases multiple scatterings are not important.[27] This assumption is generally valid if one considers that multiple scattering processes can be accounted for by adding all scattering paths that originate and terminate at the central atom (absorber). Each of these processes then behaves like $Sin(2kr_{eff})$ where $2r_{eff}$ is the total scattering path length which is much larger than that of the direct backscattering(s) from the nearest neighbors. Thus, multiple scatterings will give rise to rapidly oscillatory waves in k space which tend to cancel out. The amplitude of these waves are also significantly attenuated by the large scattering path lengths, making it relatively unimportant in comparison with the direct backscattering.[8]

On the other hand, multiple scattering in EXAFS can become important when atoms are arranged in an approximately colinear array. In such cases, the outgoing photoelectron is strongly *forward-scattered* by the intervening atom, resulting in a significant amplitude enhancement. In fact, both the amplitude and the phase are modified by the intervening atom(s) for bond angles ranging from 180° to ~75°. The effect, however, drops off very rapidly for bond angles below *ca* 150°. For these systems, it is necessary to rewrite Eq. 4 or 14 to take into account multiple scattering involving the intervening atom(s). For the sake of simplicity, let us consider only a three-atom array $A-B-C$ where A is the central atom (absorber), B is the nearest neighbor (the intervening atom), and C is the next nearest neighbor. The generalization of the following theory to more neighboring atoms is obvious. For such a system, the EXAFS of the absorber A is comprised of two contributions, one from the *backscattering* of B and the other from the scattering of C via the intermediary atom B. We shall designate these two contributions as AB and AC respectively. The former can be described quite adequately by the *backscattering* from the atom B with the *single-electron single-scattering theory* (Eq. 4 and 14). The latter, which is affected by multiple scattering involving the intervening atom B, must be treated with a generalized formulation as described below.

Let us first define a scattering angle β at atom B (and similarly γ at atom C) which is related to the bond angle $A-B-C$ (α) (*cf.* Figure 3) by

$$\beta = 180° - \alpha \qquad (20)$$

For the simple three-atom ABC system, the EXAFS contribution which corresponds to the $A-C$ peak in the Fourier transform is composed of three parts as exemplified schematically in Figure 4. The first pathway (I) is the direct backscattering from A to C and back (*viz.* $A\rightarrow C\rightarrow A$). The second pathway (II) is the multiple scattering via atom B around the triangle in either directions (*viz.* $A\rightarrow B\rightarrow C\rightarrow A$ or $A\rightarrow C\rightarrow B\rightarrow A$). This term should therefore be counted twice. The third pathway (III) is the multiple scattering via the intervening atom B in both the outgoing and incoming trip (*viz.* $A \rightarrow B \rightarrow C \rightarrow B \rightarrow A$). The EXAFS, corresponding to the $A-C$ peak, is then the sum of these three terms which reduces to[23]

$$\chi^{AC}(k) = \Omega_B(\beta,k)F_C(k)e^{-2\sigma_C^2 k^2}e^{-2r_{AC}/\lambda}\frac{Sin\left[2kr_{AC}+\phi_{AC}(k)+\omega_B(\beta,k)\right]}{k\,r_{AC}^2} \qquad (21)$$

where

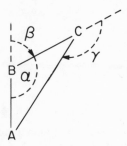

Fig. 3.   Schematic representation of a three-atom ABC system where A is the X-ray
            absorbing atom (central atom), B is the nearest neighbor, and C is the next
            nearest neighbor. Here α is the A-B-C bond angle and β and γ are scatter-
            ing angles at atom B and C, respectively.

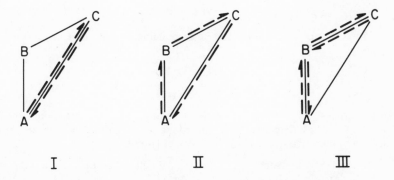

Fig. 4.   Schematic representation of three scattering pathways for a three-atom ABC
            system. Each of these pathways originates and terminates at the absorbing
            atom A. Pathway I is the direct backscattering from atom A to atom C and
            back. Pathway II is the multiple scattering via atom B around the triangle in
            either direction (only one is shown) and pathway III is the multiple scatter-
            ing via atom B in both outgoing and incoming trips.

$$\Omega_B(\beta,k) = 1 + 2\,\tilde{r}F_B(\beta,k)e^{-\Delta r/\lambda}Cos\,\tilde{\theta} + \left[\tilde{r}F_B(\beta,k)e^{-\Delta r/\lambda}\right]^2 \tag{22}$$

$$\omega_B(\beta,k) = 2\tan^{-1}\frac{\tilde{r}F_B(\beta,k)e^{-\Delta r/\lambda}Sin\,\tilde{\theta}}{1 + \tilde{r}F_B(\beta,k)e^{-\Delta r/\lambda}Cos\,\tilde{\theta}} \tag{23}$$

$$\tilde{r} = \frac{r_{AC}}{r_{AB}r_{BC}} \tag{24}$$

$$\Delta r = r_{AB} + r_{BC} - r_{AC} \tag{25}$$

$$\tilde{\theta} = \theta_B(\beta,k) + k(\Delta r) \tag{26}$$

Here $F_B(\beta,k)$ and $\theta_B(\beta,k)$ are the scattering amplitude and phase functions for the scattering angle β at atom B. Other terms in Eq. 21 have their usual meaning.

It is apparent from Eq. 21 that the effect of multiple scattering via the intervening atom B is to multiply the amplitude $F_C(k)$ by $\Omega_B(\beta,k)$ and to add $\omega_B(\beta,k)$ to the phase $\phi_{AC}(k)$. That is, if one substitutes the modified amplitude $F_C(k)\,\Omega_B(\beta,k)$ for $F_C(k)$ and the corrected phase $\phi_C(k) + \omega_B(\beta,k)$ for $\phi_C(k)$, the EXAFS data can be analyzed in the usual way. The use of multiple scattering formalism in bond angle determination will be discussed in a later section of this chapter.

The Energy Threshold Problem

It should be emphasized that the phase shifts are unique only if the energy thresholds, E_0, are specified. Changing E_0 by $\Delta E_0 = E'_0 - E_0$ will change the momentum k to

$$k' = \left[k^2 - 0.262\,(\Delta E_0)\right]^{1/2} \tag{27}$$

where k is in Å^{-1} and ΔE_0 in eV. The corresponding modification of the phase shift function will be

$$\phi'(k') = \phi(k) - 2(k' - k)r$$

$$= \phi(k) + 0.262r(\Delta E_0)/k \tag{28}$$

for $0.262(\Delta E_0) \ll k^2$. It is obvious from Eq. 28 that the difference $\Delta\phi(k) = \phi'(k') - \phi(k)$ decreases with increasing k indicating that phase shifts are more sensitive to a change in E_0 at small k than at large k.

It is clear then that, in order to fit experimental data based upon some empirical E_0 with the theoretical phase shifts, we must allow ΔE_0 to vary where $\Delta E_0 = E_0^{th} - E_0^{exp}$ with E_0^{th} and E_0^{exp} denoting the "theoretical" and "experimental" energy thresholds, respectively.

Since the determination of interatomic distance r depends on the precise knowledge of $\phi(k)$, the nonuniqueness of phase shifts naturally causes concern about the uniqueness of the distance determination. Fortunately, it can be shown that by adjusting E_0 it is not possible to produce an artificially good fit with a wrong distance r, simply because changing E_0 will affect $\phi(k)$ mainly at low k values by $\sim 0.262r(\Delta E_0)/k$ whereas changing r will affect $\phi(k)$ mostly at high k values by $2k(\Delta r)$.[8b]

The E_0 variation also helps remove the small but significant bonding effects such as electronic configurations and atomic charges to be discussed next.

The effects of valence shell electronic configuration on the amplitude as well as the scatterer and central atom phase functions are illustrated in Figures 5a-c for Cu, respectively.[28] The configurations used in the calculations are $3d^9 4s^2$ and $3d^{10} 5s^1$. It can be seen that the amplitude functions (Figure 5a) are little affected by changes in electronic configuration. The small variations at low k values are not unexpected because in this region the photoelectron energy is comparable to the valence shell binding energies. On the other hand, both the scatterer (Fig. 5b) and the absorber (Fig. 5c) phase shifts exhibit systematic variations with electronic configuration. Fortunately, the difference between various configurations diminishes as k increases. This difference, which is roughly inversely proportional to k, can largely be compensated for by changing E_0 as shown in Figure 5c (dashed curve).

The effect of atomic charge on central atom phase shift is shown in Figure 6.[28] Here we plot the theoretical ϕ_a^0, ϕ_a^1, and ϕ_a^2 for Ca and Ca^{2+}. It is immediately obvious that the dication not only has a more positive phase shift but also a larger slope. It is also readily apparent that the effect of atomic charge on phase shifts is significantly larger than that of electronic configuration. This is not unexpected since the atomic charge exerts a significant effect on the central atom potential which is experienced by both the outgoing and the incoming photoelectrons. The difference, again, can partially be compensated for by E_0 variation.

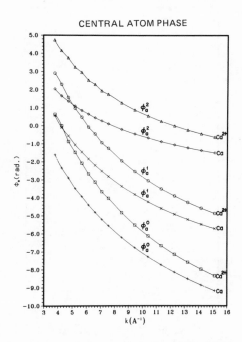

Fig. 6.   The effect of atomic charge on the central atom phase shifts as exemplified by Ca and Ca^{2+}.

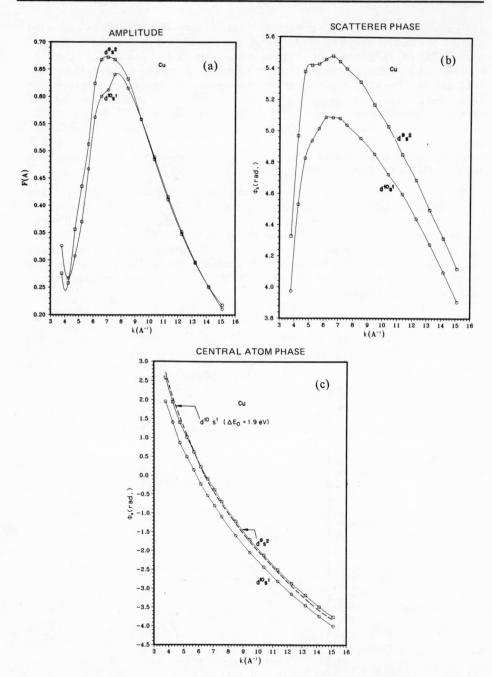

Fig. 5.   Comparisons of the amplitude (a), backscattering phase (b), and central atom phase ϕ_a^1 (c) functions for Cu with electronic configurations $3d^9 4s^2$ and $3d^{10} 4s^1$. In (c), the dashed line corresponds to $d^{10} s^1$ phase with $\Delta E_0 = 1.9$ eV.

Data Analysis

In order for EXAFS to provide accurate structural and chemical information, three assumptions must be made:[2,8,29] (1) EXAFS is a simple sum of waves due to various types of neighboring atoms (this implies that multiple scattering is relatively unimportant); (2) the amplitude function is transferable for each type of backscatterer B; and (3) the phase function is transferable for each pair of atoms $A-B$ (where A is the absorber and B is the backscatterer). Transferability implies that these functions are relatively insensitive to chemical bonding for energies 60-1000 eV above the absorption edge such that once determined for a known system, they can be applied to unknown systems which contain the same corresponding elements. The possible breakdown of the single scattering approximation, as well as its remedy, has been addressed in a previous section concerning multiple scatterings. The possible nontransferability of the amplitude functions and its causes will be dealt with in the next chapter by Stern, Bunker, and Heald.[21a] Finally, the chemical effect on the phase transferability can often be to a large extent compensated for by varying the energy threshold as discussed in the previous section. It should be emphasized that, of these three assumptions generally invoked in EXAFS data analysis, the phase transferability has proven to be the best approximation which means that EXAFS is an excellent technique for the determination of interatomic distances. The single scattering approximation works well except in cases where neighboring atoms are arranged in a linear or nearly colinear array. Finally, the amplitude transferability has been shown to be the weakest approximation due to the fact that it is highly sensitive to many variables including many-body effects, inelastic losses, and disorders. It is also sensitive to experimental conditions and the data reduction procedure (data length, cut-off range, weighting scheme, Fourier filtering, etc.). Nevertheless, amplitude transferability can still be exercised with caution for closely related systems.

There are two major approaches to EXAFS data analysis: the Fourier transform (FT) and the curve fitting (CF) techniques. In either method, the μ vs E data (Figure 1) is converted into $\chi(k)$ vs k via Eq. 2 and 3 (Figure 7a).[9] $\chi(k)$ is then multiplied by some power of k so as to compensate for the diminishing amplitudes at high k values to give $k^n\chi(k)$ for which a weighting scheme of $n = 3$ is commonly used (solid curve in Figure 7b).

The FT method[2,8] involves the Fourier transformation of $k^n\chi(k)$ in momentum (k) space over the finite k range k_{min} to k_{max} to give the radial distribution function $\rho_n(r')$ in distance (r') space (Figure 7c).

$$\rho_n(r') = \frac{1}{(2\pi)^{1/2}} \int_{k_{min}}^{k_{max}} k^n \chi(k)e^{i2kr'}dk \qquad (29)$$

Each peak in $\rho_n(r')$ is shifted from the true distance r by $\alpha = r - r'$ where α amounts to ca 0.2~0.5 Å depending upon the elements involved. α can be obtained from model compounds and transferred to the unknown systems to predict distances. For closely related systems, the approximate number of neighboring atoms can be calculated by $N = \dfrac{Ar^2}{A_s r_s^2} N_s$ where N_s, A_s, r_s and N, A, r are the number of atoms, the Fourier transform peak areas, (or peak height if the two Debye-Waller factors are similar), and the interatomic distances in the standard and the unknown compounds, respectively. This method works very well for systems with well separated peaks. The weakness of it is that since the phase functions are not exactly linear, the shift α will depend on E_0, the weighting of the data before Fourier transforming, the data range in k space, and the Debye-Waller factors.

The curve fitting (CF) technique, on the other hand, attempts to best fit (e.g. by least-squares refinements) the $k^n\chi(k)$ spectra in k space with some phenomenological models based on Eq. 4 or 14 (dashed curves in Fig. 7b and d). There are several variants of this method. The refinement procedure can be based on least-squares (commonly used) or absolute-difference minimization. One could multiply $\chi(k)$ by some power of k to give $k^n\chi(k)$ before fitting so as to emphasize different k regions (the higher the n values, the more weighting is the higher k regions) or simply apply a weighting scheme as a function of k. One could fit the entire EXAFS spectrum directly as is often done or, in cases where strong correlation exists between the amplitude and the phase, one could fit the experimental amplitude and phase functions separately *(vide infra)*. Finally, though it is possible to deal with the amplitude and the phase shifts

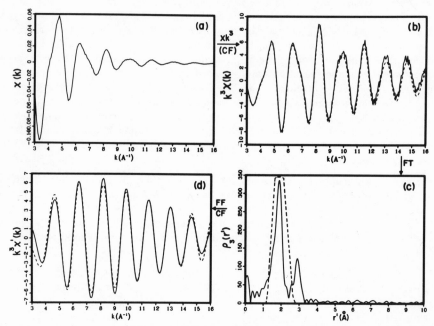

Fig. 7.   Data reduction and data analysis in EXAFS spectroscopy: (a) EXAFS spectrum $\chi(k)$ *vs* k after background removal; (b) the solid curve is the weighted EXAFS spectrum $k^3\chi(k)$ *vs* k (after multiplying $\chi(k)$ by k^3). The dashed curve represents an attempt to fit the data with a two-distance model by the curve-fitting (CF) technique; (c) Fourier transformation (FT) of the weighted EXAFS spectrum in momentum (k) space into the radial distribution function $\rho_3(r')$ *vs* r' in distance space. r' is related to the true distance r by a "phase shift" $\alpha = r - r'$. The dashed curve is the window function used to filter the major peak in Fourier-filtering (FF); (d) Fourier-filtered EXAFS spectrum $k^3\chi'(k)$ *vs* k (solid curve) of the major peak in (c) after backtransforming into k space. The dashed curve attempts to fit the filtered data with a single-distance model.

numerically, it may be convenient to parameterize these functions with simple analytical forms. The amplitude function $F(k)$ of scatterers can be parameterized with a simple sum of Lorentzians[29b]

$$F(k) = \sum_i \frac{A_i}{1 + B_i^2(k-C_i)^2} \tag{30}$$

where A is the peak height, $2/B$ is the width and C is the peak position in k space. For $Z \lesssim 36$, $36 \lesssim Z \lesssim 57$, and $57 \lesssim Z \lesssim 86$, one, two, and three Lorentzians are needed for the weighting schemes of $k^3\chi(k)$, $k^2\chi(k)$, and $k\chi(k)$, respectively. Other functional forms ranging from the simple two-parameter form[30]

$$F(k) = \frac{C}{k^\beta} \tag{31}$$

where $\beta \approx 2$ to the complicated ten-parameter form[31]

$$F(k) = \frac{a_0 + a_1k + \ldots + a_5k^5}{1 + a_6(k-a_{10}) + \ldots + a_9(k-a_{10})^4} \tag{32}$$

have also been used in the literature. The Lorentzian form is preferred in that it describes nicely the characteristic maxima of experimental amplitude curves with a minimum number of parameters. At high enough energy $(k \gg C)$, it reduces to the well-known Born approximation of scattering amplitude for fast electrons scattered elastically by a spherically symmetrical atom.[29b]

The phase functions can be parameterized by either a linear function (Eq. 33a),[2] a quadratic function (Eq. 33b),[2,29a] or a more complicated four-parameter form (Eq. 33c).[29c]

$$\phi(k) = p_0 + p_1k \tag{33a}$$

$$\phi(k) = p_0 + p_1k + p_2k^2 \tag{33b}$$

$$\phi(k) = p_0 + p_1k + p_2k^2 + a_3/k^3 \tag{33c}$$

While the central atom phase functions $\phi_a(k)$ can be fitted quite adequately with the quadratic form (Eq. 33b), the backscatterer phase functions $\phi_b(k)$ often require a functional form similar to (for $Z \lesssim 36$) or more complicated than (for $Z \gtrsim 36$) Eq. 33c.[28] In practice, since $\phi_a(k)$ has a stronger k dependence than $\phi_b(k)$, the combined phase $\phi(k)$ can often be parameterized adequately with Eq. 33b or c.

Parameterization of the amplitude and phase functions has the distinct advantage of reducing these functions into a few parameters. These parameters can first be obtained by fitting either the experimental EXAFS of structurally known model compounds or the theoretical curves and then transferred to the unknown systems for structural determination. Parameterization also provides some smoothing of the functions which, in some cases, can introduce some error.

It is obvious that the Fourier transform technique has the advantage of providing a simple physical picture — radial distribution function — of the local structure around the absorber whereas curve fitting methods can provide higher resolution and more accurate results, especially for systems with closely spaced interatomic distances. The problems of the Fourier transform technique include: (1) "side lobes" (due mainly to finite data length) which can interfere with the weaker peaks; (2) the position and shape of peaks can to some extent be affected by the weighting scheme; and (3) skewing of the transformed envelope (due to the nonlinear character of the phase and amplitude) which may affect the peak position. The problems of curve fitting methods include: (1) multi-dimensional parameter spaces which can be very time-consuming; and (2) correlations among various parameters.

A compromise of these two approaches is the Fourier filtering followed by curve fitting technique. It involves Fourier transforming the $k^n\chi(k)$ data into the distance space, selecting the distance range of interest with some smooth window (dashed curve in Fig. 7c), and back transforming the data to k space (Fig. 7d). The resulting 'filtered' EXAFS spectrum $k^n\chi'(k)$ can then be fitted with simpler models (dashed curve in Fig. 7d). This procedure has the additional advantage of simultaneous removal of the high frequency noise and the residual background as well as providing equally-spaced data points in k space. Fourier filtering, however, can cause some distortions (especially in the amplitude) at the boundaries of the spectrum. It is therefore beneficial to choose a sufficiently wide window and to use a data length shorter than the original one after filtering.

It should be mentioned that after FF, CF is not the only option. For example, for single-scatterer peaks, one may choose to decompose the EXAFS wave into phase $(\Phi(k) = 2kr + \phi(k))$ and amplitude $(A(k) = (N/kr^2)e^{-2\sigma^2k^2}e^{-2r/\lambda}F(k))$ components. These components can then be analyzed separately using either theory or model compounds. With a carefully chosen model compound (designated by the subscript s) of known and similar structure, the difference in the phase function yields

$$\Phi(k') - \Phi_s(k) = 2k'r - 2kr_s + \phi'(k') - \phi(k) . \tag{34a}$$

Eq. 34a normally does not pass through the origin due to the nonzero difference in phase $\phi'(k') - \phi(k)$ caused by the difference in E_0. By adjusting $\Delta E_0 = E_0 - E_0'$ such that $\phi'(k') - \phi(k) = 2(k - k')r$, Eq. 34a becomes

$$\Phi(k') - \Phi_s(k) = 2k(r - r_s) \tag{34b}$$

which will pass through the origin (i.e. zero intercept). In practice, therefore, a plot of $\Phi(k') - \Phi_s(k)$ vs k with the E_0 for the unknown system least-squares refined until the fitted linear curve passes through the origin gives rise to a slope of $2(r - r_s)$ from which the unknown distance r can readily be calculated. On the other hand, the amplitude envelope can be analyzed by plotting $\ln A(k)/A_s(k)$ vs k^2. A least-squares fitted linear curve will give rise to an intercept of $\ln(N/N_s \times r_s^2/r^2)$ and a slope of $2(\sigma_s^2 - \sigma^2)$, assuming other factors being equal or similar for the two systems. Knowing σ_s, N_s, r_s from the model compound as well as r from the phase analysis, it is then possible to determine the unknown σ and N from the amplitudes.[2b]

For systems with a large disparity in distances, a beat-node technique may also be used. Consider a system with one scatterer type but two sets of distances R_1 and R_2. The EXAFS can be written as

$$\chi(k) = \frac{1}{k} \tilde{A}(k) Sin[2k\tilde{R} + \tilde{\phi}(k)] \tag{35}$$

where

$$\tilde{A}(k) = A_1(k)[1 + C(k)^2 + 2C(k)Cos(2k\Delta R)]^{1/2} \tag{36}$$

$$\tilde{\phi}(k) = \phi(k) + arc\ Tan\ \frac{1 - C(k)}{1 + C(k)}\ Tan\ k\Delta R \tag{37}$$

$$A_i(k) = \frac{N_i}{R_i^2} F(k) e^{-2\sigma_i^2 k^2} e^{-2R_i/\lambda} \qquad (i = 1,2) \tag{38}$$

$$C(k) = \frac{A_2(k)}{A_1(k)} = \frac{N_2}{N_1} \frac{R_1^2}{R_2^2} e^{-2(\sigma_2^2 - \sigma_1^2)k^2} e^{-2\Delta R/\lambda} \tag{39}$$

$$\tilde{R} = (R_1 + R_2)/2 \tag{40}$$

$$\Delta R = R_1 - R_2 \tag{41}$$

Here $C(k)$ can be varied as a single parameter or fitted as a function

$$C(k) = ce^{-2Sk^2} \tag{42}$$

where $c = (N_2/N_1)(R_1^2/R_2^2)e^{-\Delta R/\lambda}$ and $S = \sigma_2^2 - \sigma_1^2$. The EXAFS now involves modified amplitude $\tilde{A}(k)$ as well as phase $\tilde{\phi}(k)$ functions which are related to the first (often the dominant) term via Equations 35-42. Knowing $F(k)$ and $\phi(k)$ from theory or model compounds, it is possible to deduce the average distance $\tilde{R}$ and the bond length difference ΔR (as well as the sign of it) along with information concerning the number of bonds and Debye Waller factors.[32]

Theoretical Amplitude and Phase Functions

It is evident that each EXAFS wave contains two sets of highly correlated variables: $\{F(k), \sigma, \lambda, N\}$ and $\{\phi(k), E_0, r\}$. Significant correlations can occur both *within* and *between* these two set of variables as well as *between* different scattering terms. In order to determine N and σ, $F(k)$ must be known reasonably well; similarly, in order to determine r, $\phi(k)$ must be known accurately.

While both $F(k)$ and $\phi(k)$ can be determined empirically from model compounds, it is clearly desirable to calculate them from first principles. By use of an electron-atom scattering model,[8b] these functions, which vary systematically with atomic number, have been calculated for most elements in the periodic table (cf. Tables I-VIII in Ref. 28). These theoretical functions[28] and their parameterized versions[29b,c] are widely used in EXAFS analyses with better than 0.5%, 10%, and 20% accuracy in r, σ, and N determinations, respectively. Some of these theoretical functions are shown in the Chapter by Lee elsewhere in this book.[33]

Bond Angle Determination

In this section, we shall describe a general method for interatomic angle determination by EXAFS on unoriented (spherically averaged) materials based on the multiple scattering theory described in a previous section.

First, let us examine how the scattering amplitude and phase functions vary with scattering angle β. Figures 8a and b depict the amplitude functions $F(\beta,k)$ for β ranging from 0° to 70° for oxygen. It is immediately apparent that the amplitude has its maximum at $\beta=0°$ and attenuates rapidly as β increases. The high k region, however, drops off much faster than the low k region. At $\beta \gtrsim 30°$, $F(\beta,k)$ is generally quite small ($\lesssim 1$) with some fine structure which changes as β varies. The complexity of the fine structure also seems to increase with increasing atomic number Z.[23]

The corresponding plots of scattering phase $\theta(\beta,k)$ vs β (Figures 9a and b) show that as β increases, the scatterer phase increases, first slowly at low β then at a faster rate. Again, at high β values ($\beta \gtrsim 30°$), complex structures tend to develop which are related to the sampling of the core levels of the scattering atoms and hence is Z (atomic number) dependent.[23]

It is the characteristic features of $F(\beta,k)$ and $\theta(\beta,k)$ as functions of β, k, and Z which form the theoretical basis for bond angle determination by EXAFS.

Next, we shall illustrate the effect of the amplitude and phase modification factors, $\Omega(\beta,k)$ and $\omega(\beta,k)$, as a function of the scattering angle β using again oxygen as an example. For clarity, we set $e^{-\Delta r/\lambda} = 1$ in Eq. 21-23. We further assume that $r_{AB} = 1.95$ Å and $r_{BC} = 1.28$ Å and calculate r_{AC} from r_{AB}, r_{BC}, and β.

Figures 10 and 11 show the results for $\Omega(\beta,k)$ and $\omega(\beta,k)$, respectively, calculated from Eq. 22-26 for $\beta=0-70°$. It is immediately obvious that multiple scattering can lead to not only amplitude enhancement ($\Omega > 1$), but also amplitude reduction ($\Omega < 1$). At $\beta \approx 0°$, $\Omega(\beta,k)$ has the maximum magnitude ($\Omega(\beta,k) \approx 9$) and is generally a flat function of k. As β increases, $\Omega(\beta,k)$ attenuates rapidly, especially in the high k region. At $\beta \approx 30°$, for example, $\Omega(\beta,k)$ drops to ~ 2.5 (amplitude enhancement) at low k values and ~ 0.4 (amplitude reduction) at high k values. For higher values of β, we find that $\Omega(\beta,k)$ exhibits characteristic features in k space for each β value, with its maxima and minima progressing smoothly as β changes. The systematic progression of these extrema as β changes stems from the composite effect of $F(\beta,k)$ and $k(\Delta r)$ as they vary systematically with β.[23]

In contrast, the β dependence of the phase modification factor $\omega(\beta,k)$ is less dramatic. At $\beta \approx 30°$ where the strongest β dependence is observed, the maximum slope of ~ -0.20 rad./Å^{-1} will give rise to a distance correction of ~ 0.10Å. For $\beta \gtrsim 50°$, $\omega(\beta,k)$ is once again a weak function of both β and k. For example, at $\beta \approx 55°$, a slope of ~ 0.10 rad./Å^{-1} will cause a correction in distance of ~ -0.05Å. It should be noted that these phase shift corrections are in general smaller than that caused by the corresponding backscatterer phase shift and much smaller than those caused by central atom phase shifts.

An examination of the three components of $\Omega(\beta,k)$ (cf. Fig. 12) revealed that at low scattering angles ($\beta \lesssim 30°$), the relative importance of the three scattering pathways follows the order: III $>>$ II $\gtrsim$ I. That is, multiple scattering via the intermediary atom B on both forward and returning trips is the dominant scattering process such that the functional form (shape) of the amplitude modification function $\Omega(\beta,k)$ resembles that of $(\tilde{r}F(\beta,k))^2$. This is the origin of the so-called focusing effect. At higher β values, the relative magnitude of the three scattering pathways follows the order: I $\gtrsim$ II $>>$ III. That is, at high scattering angles, ($\beta \gtrsim 45°$), direct backscattering (pathway I) and the "round-the-triangle" multiple scattering (pathway II) are the most important scattering processes with the latter attenuating further as β increases. $\Omega(\beta,k)$ is now determined mainly by that of $2 \tilde{r}F(\beta,k) \cos \tilde{\theta}$ (i.e. pathway II) since pathway I gives rise to only a constant 1 in Eq. 22. In short, amplitude enhancement at low scattering angles (focusing effect) stems mainly from the multiple scattering process involving the intermediary atom in both the forward and the returning trip (pathway III in Figure 4) whereas the amplitude modification (enhancement and/or attenuation) at higher scattering angles, where both the direct backscattering (pathway I) and the "round-the-triangle" multiple scattering (pathway II) are important, comes from mainly the latter

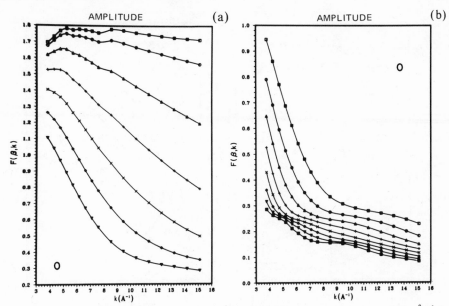

Fig. 8. Scattering amplitude $F(\beta,k)$ in Å vs photoelectron wave vector k in Å^{-1} for oxygen as a function of scattering angle β where $\beta=0°(\square)$, $5°(\bigcirc)$, $10°(\Delta)$, $15°(+)$, $20°$ (X), $25°$ $(\diamondsuit)$, $30°(\nabla)$ for (a) and $\beta=35°(\square)$, $40°(\bigcirc)$, $45°(\Delta)$, $50°(+)$, $55°$, (X) $60°$ $(\diamondsuit)$, $65°(\nabla)$, $70°(\boxtimes)$ for (b).

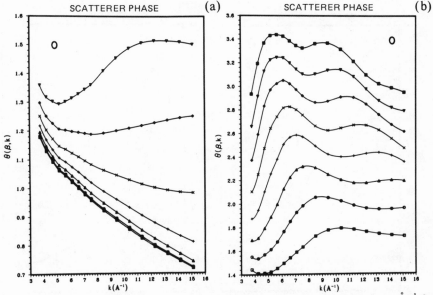

Fig. 9. Scattering phase $\theta(\beta,k)$ in radian vs photoelectron wave vector k in Å^{-1} for oxygen as a function of scattering angle β where $\beta=0°(\square)$, $5°(\bigcirc)$, $10°(\Delta)$, $15°(+)$, $20°$ (X), $25°$ $(\diamondsuit)$, $30°(\nabla)$ for (a) and $\beta=35°(\square)$, $40°(\bigcirc)$, $45°(\Delta)$, $50°(+)$, $55°$ (X), $60°$ $(\diamondsuit)$, $65°(\nabla)$, $70°(\boxtimes)$ for (b).

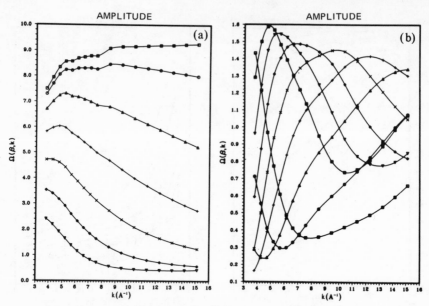

Fig. 10.   Amplitude modification factor $\Omega(\beta,k)$ *vs* photoelectron wave vector k in
            $\mathring{A}^{-1}$ for oxygen with r_{AB}=1.95Å, r_{BC}=1.28Å as a function of scattering angle
            β where β=0°(□), 5°(○), 10°(Δ), 15°(+), 20° (X), 25° (◇), 30°(▽) for (a)
            and β=35°(□), 40°(○), 45°(Δ), 50°(+), 55° (X), 60° (◇), 65°(▽), 70°(⊠)
            for (b).

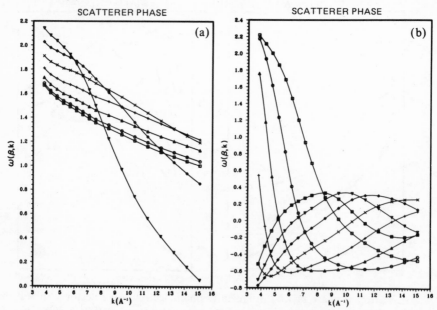

Fig. 11.   Phase modification factor $\omega(\beta,k)$ in radian *vs* photoelectron wave vector k
            in $\mathring{A}^{-1}$ for oxygen with r_{AB}=1.95Å, r_{BC}=1.28Å as a function of scattering
            angle β where β=0°(□), 5°(○), 10°(Δ), 15°(+), 20° (X), 25° (◇), 30°(▽),
            for (a) and β=35°(□), 40°(○), 45°(Δ), 50°(+), 55° (X), 60° (◇), 65°(▽),
            70°(⊠) for (b).

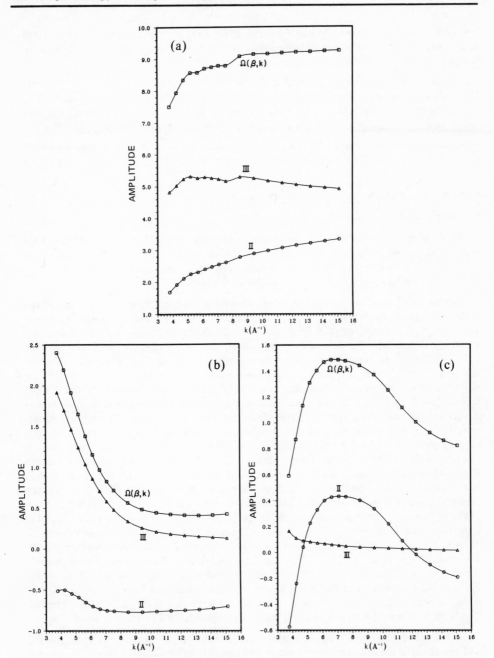

Fig. 12. Multiple scattering contributions to the amplitude modification factor $\Omega(\beta,k)(\square)$ *vs* k in Å^{-1} due to pathways II ($\triangle$) and III ($\bigcirc$) in Figure 2 for oxygen at different scattering angles β where $\beta=0°(a)$, $30°(b)$, and $60°(c)$. Curves II and III correspond to the functions $2\tilde{r}F(\beta,k)Cos\,\tilde{\theta}(\beta,k)$ and $[\tilde{r}F(\beta,k)]^2$, respectively, defined in the text.

multiple scattering process. While focusing effect can, at low scattering angles (large bond angles), cause an order of magnitude enhancement in amplitude, its impact on the phase is, fortunately, less dramatic.

To determine the bond angle A-B-C ($= 180° - \beta$), the distance r_{AC}, and the Debye-Waller factor σ_C, one simply attempts to best fit the experimental curve with Eq. 21 by least-squares refining β, r_{AC}, σ_C, ΔE_{o_C}, η_C, and S_C. The distance r_{AB} and the energy correction ΔE_{o_B} are taken from the AB fit and held constant throughout the curve fitting process. The distance r_{BC} can be calculated from r_{AB}, r_{BC}, and β in each cycle of refinement. The accuracy for bond angle determinations is better than 5% or ca 5°.[23]

EXPERIMENTATION

Synchrotron Radiation Sources

Most of recent EXAFS work are done using synchrotron radiation. These highly collimated, plane polarized, and precisely pulsed X-ray sources, with fluxes of $10^{12} \sim 10^{16}$ photons/sec/mrad/mA, greatly improve the signal-to-noise ratio by $10^3 \sim 10^6$ over that obtainable from conventional sources.

Synchrotron radiation is emitted when relativistic charged particles travel in curved paths in magnetic fields. It is a natural by-product (in fact, a nuisance due to the power loss) of high energy physics experiments in which electrons and positrons are accelerated to nearly the velocity of light in an evacuated chamber of a storage ring before colliding to form new elementary particles. Hence, synchrotron radiation research is traditionally operated in a parasitic mode, though new dedicated beams are now available. At present, there are five synchrotron radiation sources in the United States. SSRL (Stanford Synchrotron Radiation Laboratory) at Stanford, which extracts its synchrotron radiation from SPEAR (Stanford Positron Electron Accelerator Ring), operates in a 50% dedicated and 50% parasitic mode supplying synchrotron radiations ranging from UV to hard X-rays. CHESS (Cornell High Energy Synchrotron Source) at Cornell, which gets its high energy photons from CESR (Cornell Electron Storage Ring), operates in parasitic mode, providing high flux hard X-rays ideal for high energy X-ray experiments. NSLS (National Synchrotron Light Source) at BNL (Brookhaven National Laboratory), which is still under construction and scheduled for completion in 1981, will have two storage rings providing dedicated synchrotron radiations in both the ultraviolet and high energy X-ray region. These three major U.S. facilities will be described in detail by other authors in three later chapters.[34-36] Other synchrotron radiation facilities include, in U.S.A., Tantalus I and Aladdin at the University of Wisconsin which are dedicated to the production of UV and soft X-rays and SURF II at the National Bureau of Standards for low-energy vacuum-ultraviolet light as well as DORIS, which is part of DESY at Hamburg, Germany; ARUS in USSR; BONN in Germany; INS-SOR II in Japan; Frascati in Italy; NINA in England; DCI in France; etc.

The spectral distribution of the emitted synchrotron radiation depends upon the size and the running conditions (beam energy and current) of the storage ring. The characteristics of the emitted synchrotron radiations under various running conditions at the three major centers in the United States can be found in three later Chapters by Bienenstock,[34] Batterman,[35] and Hastings.[36] It suffices to say here that the larger the ring and the higher the beam energy and current, the higher will be the energy and intensity of the emitted synchrotron radiation.

It is important to choose an absorption edge of a particular absorber whose energy corresponds to the brightest synchrotron radiation in the spectral distribution as well as to minimal harmonics in the X-ray beam. Currently at SSRL, the EXAFS I and wiggler beam lines cover an energy range of 3-35 KeV (which includes the K-edge of elements from potassium to iodine and the L-edges of elements through the end of the periodic table); the focused beam line, which has ca 100 times greater relative intensity, spans a 3-9 KeV range (which include the K-edge of elements from potassium to copper). The wiggler line (using wiggler magnets) not only provides a greater X-ray intensity (by a factor of 6 $\sim$ 100) than the EXAFS I line (using bending magnets), but also extends to 2-10 times higher X-ray energies.[34]

In short, the advantages of synchrotron radiations are: (1) high intensity (10^3 times over characteristic lines and 10^6 times over white lines of conventional X-ray tubes); (2) tunability over a wide energy range with a continuous spectrum; (3) high collimation; (4) plane-polarization, which is useful in single-crystal studies; and (5) precisely pulsed time structure (nanosecond pulses separated by micro- or milliseconds), which is important in time dependent studies.

Other Sources

A new source of high-intensity, nanosecond-pulsed *soft* X-rays ($\leqslant 3$ KeV) has recently been generated from a laser-produced plasma.[37] This technique is useful for EXAFS studies of low Z elements (K-edge of the elements from carbon to sulfur and L-edge of the elements from sulfur to molybdenum) which are at present difficult to study with other sources. It enables studies of transient species since an EXAFS spectrum can be recorded on film in a few nanoseconds with a *single pulse* of laser-produced X-rays.

X-ray sources using the continuous bremsstrahlung output from a rotating anode X-ray tube can also be used in EXAFS studies.[38] Since the intensity is ca 10^3 times weaker than the synchrotron source, each spectrum generally takes 2-8 hours to achieve the desirable S/N ratio. Furthermore, the radiation from an X-ray tube contains characteristic lines from the target material and/or contaminants which may interfere with the EXAFS measurements. On the other hand, the conventional sources have the distinct advantage of being easily accessible and continuously available. In this respect, these laboratory-scale conventional sources are very useful for stable samples with high concentrations.

Sample Considerations

Selecting a particular absorption edge of a specific element in the sample requires some consideration. First, $K(1s)$ and $L_I(2s)$ edges require the same set of amplitude and phase functions, whereas an $L_{II}(2p_{1/2})$ or $L_{III}(2p_{3/2})$ edge requires a different set of phase shifts.[28] Second, the energy range above and below the edge must be >600 eV to prevent interference from any adjacent edge(s). Third, the edge jump must be sizable since EXAFS is a measure of the fractional oscillation of the absorption relative to the edge jump. Fourth, absorption (or absorption edges) due to other elements in the material should be minimized or avoided to prevent a sloping background (or interference). Finally, at low energies, absorptions due to sample cells and intervening air may pose some difficulties. The former problem can be reduced by the use of X-ray transparent tapes as windows whereas the latter may require a helium path.

The sample thickness or concentration depends on the total absorption coefficient of the material and the detection technique to be discussed in the next section. For transmission experiments, optimum signal-to-noise ratio occurs at $\mu x \approx 2$. For concentrated samples with high absorption coefficient, it may be necessary to dilute the sample with low Z inert materials such as boron nitride or sucrose in order to attain a manageable thickness. On the other hand, for dilute samples ($\sim 10^{-3}$M), it may be necessary to use the fluorescence technique and/or to average a number (n) of spectra to increase the S/N ratio (by $n^{1/2}$). The sample thickness must be fairly constant in the area of the beam.

Sample homogeneity is of utmost importance in the sample preparation, especially for transmission experiments involving low Z absorbers *(vide infra)*. Pinholes or inhomogeneity in the sample can cause undesirable fluctuation in absorption across the sample.

Higher harmonics in the incident beam, which can introduce serious error in the measurement since higher energy harmonics can pass through the sample basically unattenuated, can be eliminated by detuning the monochromator slightly and/or by changing the sample thickness.

Finally, for weak chemical bonds or large σ_{vib}, it may be necessary to perform the measurements at lower temperatures.

Experimental Techniques

There are several well-developed modes of EXAFS measurements, most of which use photons as the primary excitation source. They differ mainly in the detection technique: some involve photons (transmission and fluorescence) whereas others involve electrons (Auger or partial electron yields). These experimental techniques can also be broadly classified as direct and indirect measurements. The transmission technique measures the transmitted X-ray intensity directly. On the other hand, the fluorescence technique measures the secondary fluorescent photons while the Auger or partial electron yield techniques measure the secondary electrons. A recently developed method of EXAFS measurement, Electron Energy Loss Spectroscopy (EELS), involves electrons in both the excitation and the detection modes using either an electron spectrometer or a transmission electron microscope. We shall describe each of these methods separately in the following sections.

Transmission

Figure 13 depicts schematically an experimental setup for EXAFS measurements in the transmission mode.[3a] This is the original design for EXAFS I at SSRL. The synchrotron radiation from the storage ring (indicated disproportionally by a small circle at the right) is monochromatized with either a double crystal or a curved mirror monochromator in a helium-filled chamber with beryllium windows. The configuration of a channel-cut crystal or a parallel double-crystal monochromator is shown in Fig. 14. For a selected Bragg spacing d, the central wavelength of the output beam will be

$$\lambda = 2d_{hkl}\ Sin\ \theta_B \tag{43}$$

when the input beam makes an angle θ_B with the Bragg planes (i.e., satisfying the Bragg reflection condition). The output beam is displaced from the input beam by $h = 2D\ Cos\ \theta_B$ where D is the spacing (typically 1 cm) between the two crystals and

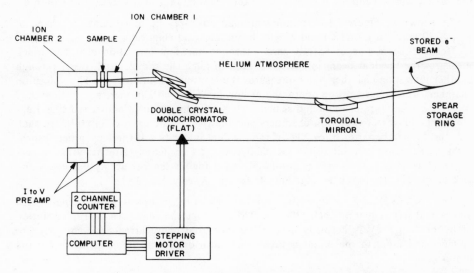

Fig. 13.   Schematic diagram of the EXAFS spectrometer in transmission mode. The synchrotron radiation from the electron storage ring, represented disproportionately by a broken ellipse at the upper right corner, is collected by a toroidal mirror and monochromatized by a double crystal monochromator. The incident beam intensity is measured by the ionization chamber 1 and the transmitted beam intensity by the ionization chamber 2. The energy of the beam is changed by changing the angle between the monochromator crystal and the incident beam.

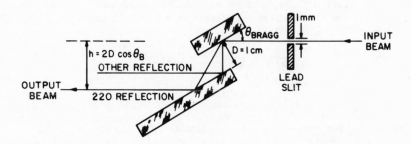

Fig. 14.   Schematic diagram of a channel cut crystal monochromator (see text).

θ_B is the Bragg angle for the chosen reflection. The parallel crystal configuration ensures that the output and input beams propagate in the same direction. Furthermore, the sample can be centered with respect to the beam by the small vertical translation h.

As shown in Figure 13, the monochromatized X-ray beam (rectangular shape of $\sim 1 \times 20$ mm^2 for EXAFS I and wiggler beam lines and elliptical spot of $\sim 2 \times 3$ mm^2 for focused beam line at SSRL) passes through the first ionization chamber which measures the incident beam intensity I_0, then through the sample, and finally through another ionization chamber which measures the transmitted intensity I. The ionization chambers which measure the photoion current, are filled with an inert gas, such as He, N_2, Ar, or a mixture of them and sealed with X-ray transparent windows (such as Kapton). The length of these ion chambers and the type of the gaseous mixture are such that the I_0 chamber absorbs ca 20% of the incident intensity and the I chamber absorbs most of the transmitted intensity. Generally, I_0 and I are 6 and 12 inches long, respectively. Low Z gas mixtures are commonly used for low energy absorption edges and vice versa (e.g., N_2 for Fe K edge and Ar for Mo K edge EXAFS).

The EXAFS spectra are typically recorded with an integration time of 2 s/point with $400 \sim 600$ steps covering about $800 \sim 1000$ eV above the edge. An energy resolution of better than 5 eV is normally used. The measured photocurrents are converted to voltages and input to the minicomputer which also drives the stepping motor for the monochromator.

For 'concentrated' samples, transmission mode is the preferred technique since it provides the best signal-to-noise ratio. For thick, uniform samples containing absorber A and other constituents B, the signal-to-noise (S/N) ratio can be shown to be[39]

$$S/N = \mu_A X e^{-\mu_T X/2} \left[\frac{\Delta \mu_A}{\mu_A} I_0^{1/2} \right] \tag{44}$$

Here I_0 is the intensity of the incident beam, X is the sample thickness, $\Delta\mu_A/\mu_A$ is the EXAFS signal, and the absorption coefficients (all functions of E and in cm^{-1}) are defined as follows

$$\mu_T = \mu_A + \mu_B \tag{45a}$$
$$\mu_A = N_A \sigma_A \tag{45b}$$
$$\mu_B = N_B \sigma_B = \sum_{i \neq A} N_i \sigma_i \tag{45c}$$

where N and σ are the concentration (number of atoms/cm^3) and the absorption cross-section (in cm^2), respectively. Differentiating S/N with respect to X, we obtain the maximum S/N at $\mu_T X = 2$ and

$$S/N = 2e^{-1} \left[\frac{\Delta\mu_A}{\mu_T} I_0^{1/2} \right] \tag{46a}$$
$$= 0.736 \frac{\mu_A}{\mu_T} \left[\frac{\Delta\mu_A}{\mu_A} I_0^{1/2} \right] \tag{46b}$$

If we also include the noise in the I_0 chamber (with absorption coefficient μ and length x), the optimal S/N ratio occurs at $\mu_T X = 2.56$ for the sample and $(\mu x)_{I_0} = 0.25$ for the I_0 chamber, and Eq. 46b becomes

$$S/N = 0.557 \frac{\mu_A}{\mu_T} \left[\frac{\Delta\mu_A}{\mu_A} I_0^{1/2} \right] . \tag{47}$$

To attain a S/N ratio of better than 1:1 in 1 second with a typical $I_0 \sim 10^{11}$ photons/sec (from storage ring) and $\Delta\mu_A/\mu_A \sim 10^{-2}$,

$$\frac{\mu_A}{\mu_T} = \frac{N_A \sigma_A}{N_B \sigma_B} \gtrsim 10^{-3} \tag{48}$$

Obviously the concentration required depends on the ratio σ_A/σ_B. For a high Z element in a low Z host material (or solvent), where $\sigma_A > \sigma_B$ since σ is roughly proportional to Z^4, a lower concentration (or equivalently, N_A/N_B) is required than in the reverse case. For example, for $\sigma_A/\sigma_B \sim 10^2$ which is commonly the case for biological systems containing heavy metals (e.g. iron proteins), $N_A/N_B \sim 10^{-5}$ which is in the millimolar region (1 mM aqueous solution corresponds to ca $N_A/N_B \sim 2 \times 10^{-5}$). On the other hand, for $\sigma_A/\sigma_B \sim 1$ as is the case in calcium proteins, $N_A/N_B \sim 10^{-3}$, a concentration of ~ 100 mM is required to achieve $S/N \sim 1$. In short, systems with high Z atoms in a low Z medium are always easier to measure.

For a thin sample of absorber A with thickness X on a substrate B with thickness X_B and absorption coefficient μ_B,

$$S/N = e^{-\mu_B X_B/2} \, \mu_A X \left[\frac{\Delta\mu_A}{\mu_A} I_0^{1/2} \right]. \tag{49}$$

It should be cautioned that other systematic effects such as beam instability, beam positioning with respect to the sample, sample inhomogeneity and thickness variation, etc., can severely worsen the signal-to-noise ratio.

Fluorescence

The fluorescence technique makes use of the fact that an excited absorber atom with a core hole (inner shell vacancy) can relax by undergoing a radiative transition from a higher electronic shell to the vacancy, thereby producing X-ray photons whose energy correspond to the energy difference between the two shells. For example, a K-shell vacancy in an iron atom produced by photoionization with an X-ray photon of 7.112 keV can return to a ground state by emitting X-ray photons K_{α_1}, K_{α_2}, K_{β_1}, ... with energies of 6.403, 6.390, 7.057, ... keV and intensity ratios of 1:0.5:0.167... The K_α, K_β, and K_γ lines, in descending order of intensity, correspond to electronic transitions from the L, M, and N shells to the K-shell, respectively.

The fluorescent yield, or the radiative probability, ϵ_f is a monotonically increasing function of atomic number Z, and is larger for K line emissions than L line emissions. For example, the K-shell fluorescent yield for O, Cl, Fe, Mo, and Pt are 0.0058, 0.094, 0.347, 0.764, and 0.963, respectively. Basically, for $Z > 50$, ϵ_f approaches 1. For comparison, the L-shell fluorescent yields of Mo and Pt are 0.067 and 0.33, respectively. Since the fluorescent yield is nearly independent of excitation energy far above the threshold energy (but may vary somewhat near the absorption edge), it is a direct measure of the absorption rate of the X-ray absorbing atoms.

The setup for a fluorescence EXAFS experiment[7a] is analogous to that of the transmission mode with the exception that, instead of measuring the transmitted intensity I, the fluorescence X-ray intensity I_f is measured. Though X-ray fluorescence is isotropic, it is often advantageous to orient the sample such that it makes an angle of 45° with both the incident beam and the fluorescence radiation detector so as to minimize interference from the incident beam and other scattered or reflected beams. A scintillation counter or a non-dispersive solid-state detector can be used as the detector. An X-ray filter assembly has been designed which not only filters out the unwanted elastic or inelastic scatterings, but also minimize the fluorescence background from the filter.[7c] For K-edge EXAFS of an element with atomic number Z, the element with atomic number $(Z - 1)$ can be used as the filter. This development extends the detection limit to lower concentrations. More recently, detectors using a combination of a curved crystal analyzer and a scintillation counter have been designed for fluorescence experiments on very dilute systems.[40,41] The limitation of these latter detectors is that each detector is good only for a limited energy range.

Discrimination against unwanted radiations is commonly done on the basis of energy. The resolutions for scintillation, solid-state, and crystal detectors at 8 keV are roughly 2000, 200, and 20 eV, respectively. The corresponding limiting counting rates are approximately 400,000, 40,000, and essentially unlimited counts/sec for each of these detectors. Finally, the solid angle acceptance for these detectors can be increased substantially by using an assembly of several individual detectors. For example, an array of 3×3 scintillation counters has been used in fluorescence experiments.

For 'dilute' samples such as biological and impurity systems, background absorption due to other atoms may produce a large slope in μ vs E. That is, the absorption μ_A due to the EXAFS-producing element A (the absorber) accounts for only a small part of the total absorption μ_T. For such systems, the transmission method is not suitable since the X-ray intensity will be substantially attenuated by the background material and it will be very difficult to detect the small EXAFS superimposed on the slopping background. For such systems, the fluorescence technique is preferred in that it detects the fluorescence of the absorber directly, thereby avoiding the absorption due to other constituents in the sample.

The S/N ratio for the fluorescence experiment has been shown to be (ignoring fluctuation in the I_0 ion chamber since the fluorescence intensity is generally much smaller than the incident beam intensity)[7a,39]

$$ S/N = \sqrt{\frac{(\Omega/4\pi)\epsilon_f \mu_A}{\mu_T + \mu_T{}'}} \left[\frac{\Delta\mu_A}{\mu_A} I_0^{1/2} \right] \tag{50} $$

where ϵ_f is the fluorescent yield and $\Omega/4\pi$ is the solid angle acceptance for the detector. Here μ_T and $\mu_T{}'$ denote the total absorption coefficients at E, the incident beam energy, and E_f, the fluorescent energy, respectively. To achieve a $S/N \sim 1$ and assuming $\Omega/4\pi \sim 10^{-2}$, $\epsilon_f \sim 0.1 - 1$, $\Delta\mu_A/\mu_A \sim 10^{-2}$, and $I_0 \sim 10^{11}$ photons/sec, $\mu_A/\mu_T \approx 10^{-5} - 10^{-4}$. This is one or two orders of magnitude lower than that for transmission experiments under similar conditions. For $\sigma_A/\sigma_B \sim 10^2$, $N_A/N_B \approx 10^{-7} - 10^{-6}$ which is in the submillimolar region. Thus, fluorescence technique can provide more sensitivity for dilute systems than the transmission method. For concentrated samples, on the other hand, the transmission technique provide better signal-to-noise ratios.

The background radiation I_b arising from quasi-elastic and inelastic scatterings can largely be reduced by the use of an absorption filter, a crystal detector, or by energy discrimination of a scintillation detector. For very dilute systems where I_b approaches I_f, the S/N ratio deteriorates to

$$S/N = \sqrt{\frac{(\Omega/4\pi)\epsilon_f \, \mu_A}{(1 + I_b/I_f)(\mu_T + \mu_T')} \left[\frac{\Delta\mu_A}{\mu_A} I_0^{1/2}\right]}$$ (51)

For a thin sample of A with thickness X on a substrate B,

$$S/N = \sqrt{\frac{(\Omega/4\pi)\epsilon_f \, \mu_A X}{1 + I_b/I_f} \left[\frac{\Delta\mu_A}{\mu_A} I_0^{1/2}\right]}$$ (52)

where I_b is the background contribution from the substrate.

SEXAFS

Two newly developed EXAFS techniques involve the measurements in vacuum of either the Auger[7d] or secondary electrons[7e] produced during the relaxation of an atom that has absorbed a photon. Since the penetration depth for electrons is ~ 20 Å, these methods are particularly useful for surface EXAFS (SEXAFS) of thin films of a few atomic layers. This method largely eliminates the background due to the substrate.

The experimental setup is similar to that of the fluorescence experiment except that it requires a high vacuum chamber and an electron detector. The detector can be either an energy analyzer, as in the case of Auger measurement, or an electron multiplier, as in the case of total or partial secondary electron yield. The energy analyzer gives rise to better resolution while the electron multiplier has larger solid angle acceptance.

The equations governing the S/N ratio for these experiments[39] are similar to that of the fluorescence technique except that the fluorescent yield ϵ_f is replaced by the nonradiative yield $\epsilon_n = 1 - \epsilon_f$, μ_T' (which is a function of E_f) by n (which is a function of E), and I_f by I_n:

$$S/N = \sqrt{\frac{(\Omega/4\pi)\epsilon_n \mu_A}{(1 + I_b/I_n)(\mu_T + n)} \left[\frac{\Delta\mu_A}{\mu_A} I_0^{1/2}\right]}$$ (53)

The exponential attenuation of the electrons is, approximately,

$$n(E) = 2 \times 10^8 [E\,(eV)]^{-1/2} \, cm^{-1} .$$ (54)

For $E \sim 1000$ eV, $n(E) \sim 10^7$ cm^{-1}, the penetration depth is $1/n(E) \sim 10$ Å.

For a thin film of absorber A of thickness X on substrate B,

$$S/N = \sqrt{\frac{(\Omega/4\pi)\epsilon_n \mu_A X}{1 + I_b/I_n} \left[\frac{\Delta\mu_A}{\mu_A} I_0^{1/2}\right]}$$ (55)

where I_b is the background contribution from the substrate and I_n is the nonradiative signal (both in electron counts/sec). Assuming $I_b/I_n \ll 1$, $(\Omega/4\pi) \sim 10^{-2}$, $\epsilon_n \sim 0.1$, $\Delta\mu_A/\mu_A \sim 10^{-2}$, and $I_0 \sim 10^{11}$ photons/sec, $\mu_A X \approx 10^{-4}$ in order to achieve a S/N ratio of 1. For $\mu_A \sim 1000$ cm^{-1}, $X \approx 10$ Å. For low Z elements where ϵ_n approaches 1, X approaches a few Å. It is therefore possible to study a monolayer or submonolayer amount of low Z or intermediate Z (note that ϵ_n decreases with Z) elements with these surface-sensitive EXAFS techniques. The advantage of this technique is that it largely removes the background contribution for more than monolayer coverage (more so than the fluorescence technique) since the penetration depth for electrons ($\lesssim 20$ Å) is much less than that for photons.

Electron Energy Loss Spectroscopy (EELS)

All the EXAFS experimental techniques discussed thus far employ photons as the primary excitation source. High energy electrons ($\sim$ 200 keV) can also be used as the excitation source. By measuring the momentum change or the energy loss of incident electrons, EXAFS information can be obtained. More specifically, when high energy electrons are scattered inelastically from a thin sample, the energy loss is a direct measure of the electronic excitation spectrum.[7f−j]

This type of inelastic electron scattering experiments can be performed with either a high vacuum electron spectrometer[7i] or a transmission electron microscope (TEM or STEM),[42] both must be equipped with an electron energy analyzer. This mode of EXAFS experimentation is particularly useful for low Z elements (up to ca 1000 eV in edge energy), and thin film samples (50 $\sim$ 500 Å). The beam diameter is $\sim$ 250 μm for the electron spectrometer and 1 $\sim$ 10 nm for the electron microscope. In the case of electron microscopy, the small beam size gives rise to the distinct possibility of EXAFS structural characterization of spatially resolved, microscopic areas of very small and thin samples.[43,44] These two modes of EXAFS measurements will be discussed in detail by others later in this book.

Applications

In this section, applications of EXAFS to biological and inorganic systems will be discussed in order to illustrate the usefulness as well as the limitations of the technique. The use of EXAFS in materials research will be described by other authors in later chapters.

Biochemical Systems

EXAFS can be used to probe the prosthetic group of a protein, thereby allowing structure-function correlation or assessment of steric *vs* electronic effects of the active site. This is exemplified by our EXAFS studies of iron-sulfur proteins.[45] It is known that there are four prototypes of non-heme iron-sulfur proteins containing one, two, four or eight iron atoms. The minimal prosthetic groups are $Fe(SR)_4$ in rubredoxin, $Fe_2S_2(SR)_4$ in plant ferredoxins (Fd), and $Fe_4S_4(SR)_4$ in 'high-potential' iron proteins (HIPIP) and bacterial ferredoxins (*cf.* Fig. 15a-c). Iron EXAFS of these proteins are dominated by neighboring sulfur and iron atoms of the active sites.[46,47] The amplitude envelope of the monomeric $Fe(SR)_4$ species varies smoothly with k (*cf.* Figure 15g), indicative of a single-shell system with one type of distance. In contrast, the amplitude envelope of the diiron and tetrairon oligomers exhibits a 'beat' node at $k \approx 7$ Å^{-1} which is characteristic of two-shell systems with two types of distances (*cf.* Figure 15h,i). The frequency at lower k region, which reflects the shorter $Fe-S$ bonds, is lower than the frequency at larger k region, which is indicative of the longer $Fe-Fe$ bonds. Fourier transforms of the $\chi(k)k^3$ data reveal in both solid and solution states only one peak for the monomer (*cf.* Figure 15d) but two peaks for the two oligomers (*cf.* Figure 15e,f). The major peak can be assigned to the $Fe-S$ bonds while the minor peak at a larger distance in each of the oligomers can be assigned to the $Fe-Fe$ bonds.[46,47]

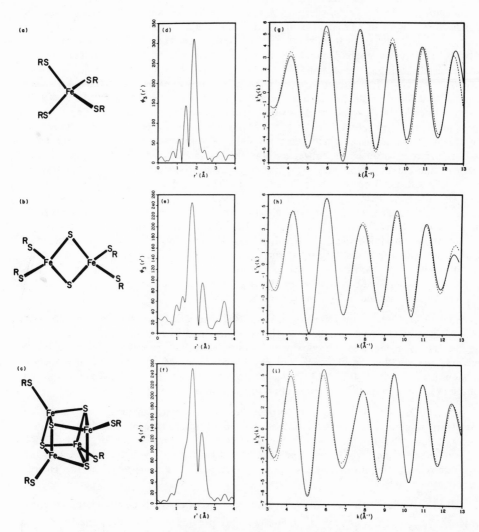

Fig. 15.   Left: the three prototypes of iron-sulfur protein active sites: (a) monomer; (b) dimer, and (c) tetramer. Middle: the Fourier transforms of the corresponding EXAFS spectra ($k = 3 \sim 14 \text{ Å}^{-1}$). Right: the corresponding Fourier filtered (window: $r' = 0.9 \sim 3.5\text{Å}$) EXAFS spectra (solid curves) and the theoretical fits (dashed curves). The major peak in each case is assigned to the $Fe-S$ distances. The minor peak at a larger distance for the oligomers are $Fe-Fe$ distances. The minor peaks to the left of the $Fe-S$ peak are due to residual background and/or Fourier truncation.

In most applications, EXAFS provides only the average distances. Nevertheless, the Debye-Waller factor can indicate the spread of the distances. A case in point is the rubredoxin problem. Protein crystallography initially revealed two kinds of $Fe-S$ bonds consisting of three distances of normal bond lengths at $r_3 = 2.30$(av.) Å and one unusually short distance at $r_1 = 2.05(3)$ Å.[48] Based on the Debye-Waller factor of $0.049(15)$ Å from a single-distance fit and a reasonable σ_{vib} of 0.045 Å calculated from an $Fe-S$ stretching frequency of 314 cm^{-1}, we concluded that the four $Fe-S$ bonds (average $2.27(1)$ Å) in rubredoxin are chemically equivalent to within 0.04 Å (with a limit of $+0.06$ or -0.04 Å) for either the 3:1 or the 2:2 ($m{:}n$) model (cf. Eq. 6,9).[46] On the other hand, if we fit the data with a two-distance model with three $Fe-S$ bonds at r_3 and one $Fe-S$ bond at r_1, a broad least-squares residuals minimum (curve A), shown in Figure 16, is obtained when we allow all parameters to vary. The reason is that strong correlation exists between σ and the distance spread Δr. In order to better estimate the difference $\Delta r = r_1 - r_3$, one must fix σ at some reasonable value such as σ_{vib} (curves B, C, D correspond to fixing σ at 0.030, 0.045 (σ_{vib}) and 0.049 Å, respectively). In fact, comparisons with model structures suggest that such estimates represent an upper limit of the true disparity in distances.[46] If the distance spread is great enough to produce a beat node in the EXAFS amplitude, the correlation between σ and Δr will diminish such that it will be possible to resolve the individual distances. Consistent with the EXAFS results, further crystallographic refinements have revised the distances such that $r_3 - r_1 = 0.10$ Å. It is now clear that the four $Fe-S$ bonds in rubredoxin in both solid and solution states are chemically equivalent. Similar conclusions were also reached by Stern and coworkers.[49]

EXAFS provides an excellent opportunity for correlating structural parameters with redox states of the proteins, both in the *solid state* and in *solution*. For example, the average $Fe-S$ distance in rubredoxin was found to lengthen by 0.06Å upon reduction[46] whereas the lengthening of $Fe-S$ and $Fe-Fe$ bonds in the oligomers upon reduction was quite small (viz., ≤ 0.03Å).[47] Furthermore, little change (within 0.02 Å) was found in the average structural parameters upon dissolution of the proteins.[47]

EXAFS has also been used to probe the stereochemical nature of a variety of metal-containing biopolymers. For the presumably relaxed porphyrin rings in deoxyhemoglobins, the iron atom was found to be only 0.2 Å (with a limit of $+0.1$ or -0.2 Å) out of the plane.[50a,b] This is significantly less than the 0.75 Å displacement of the iron atom from the mean tetranitrogen porphyrin plane as suggested by the Hoard-Perutz model.[51a,b]

EXAFS studies[52] on carbonic anhydrase, an enzyme which catalyzes the reaction $H_2O + CO_2 \underset{\leftarrow}{\rightarrow} H_2CO_3$, and on its iodide complex provide evidence that the iodide, an anionic inhibitor, is directly coordinated to the zinc atom at a normal covalent $Zn-I$ distance of $2.65(6)$ Å. This should be contrasted with the crystallographic finding that the iodide ion lies between 3.5 and 3.7 Å from the zinc atom.

An elegant study[53] on the iron-storage protein ferritin indicated that each of the iron atoms in the micellar core is surrounded by $6.4(6)$ oxygens at $1.95(2)$Å (probably in a distorted octahedral arrangement) with $7(1)$ neighboring iron atoms at $3.29(5)$Å. A layered, strip structure, in which the iron atoms are in the interstices between two nearly close-packed layers of oxygens with approximate sixfold rotational symmetry, was proposed. These authors believe that the ferritin core is formed by a folding of one such strip back and forth upon itself in the form of a pleat.

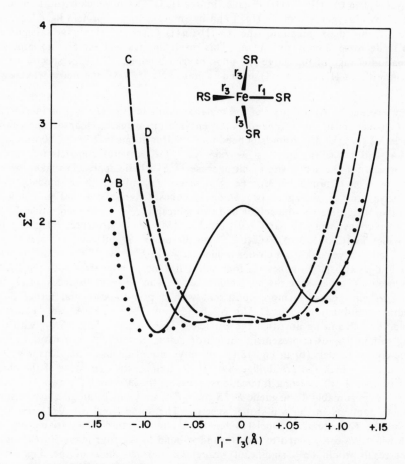

Fig. 16.   The chi-squares minimization of the curve fitting of Fourier filtered EXAFS data of rubredoxin with a *single-shell two-distance* model. Σ^2, the sum of squares of the least-squares residuals, is plotted as a function of $(r_1 - r_3)$ where r_1 and r_3 are the distances of one and three $Fe-S$ bonds, respectively. The outer curve A(.....) was calculated by fixing values of $r_1 - r_3$ and varying the 4 parameters $\frac{1}{4}(r_1 + 3r_3)$, the Debye-Waller factor σ, the scale factor, and the energy threshold E_0. The other three curves (B, C, and D) show the effect of fixing σ at 0.030 (—), 0.045 (----), and 0.049 (-.-.-.-) Å while allowing the remaining 3 parameters to vary.

A recent EXAFS study[54] on oxy- and deoxyhemocyanin (Hc) from *Busycon canaliculatum* led to a model of the Hc binding site as having two *Cu* atoms bound to the protein via three histidine ligands at average distances of 1.96 and 1.95 Å (first-shell), respectively. The *Cu* (II)-*Cu* (II) distance in oxy Hc is 3.67 Å which is short enough to be bridged by dioxygen (formally O_2^{2-}) and by an atom from a protein ligand, possibly tyrosine. Upon deoxygenation, the *Cu* (I)-*Cu* (I) distance (tentative assignment) appears to decrease slightly to 3.39 Å. This result implies that there is no large-scale conformation change of the active site upon oxygen binding. The edge spectra are also consistent with *Cu* (II) and *Cu* (I) bound to imidazole in oxy- and deoxy Hc, respectively.

From a recent EXAFS study[55] of the molybdenum site (the *MoFe* protein as well as the *FeMo* cofactor) in the nitrogen fixation enzyme nitrogenase, Hodgson and coworkers concluded that the *Mo* atom is bonded to 3.8 sulfur atoms at 2.35 Å, 3.0 iron atoms at 2.72 Å, and 1-2 sulfur atoms at ~ 2.55 Å.[55a] Two structural models (*cf.* Fig. 17) were proposed for the active site of nitrogenase.[55b] Synthetic approaches taken by various groups[56] have produced *Mo*−*Fe*−*S* clusters resembling these models. They include structures consisting of two *MoFe*$_3$*S*$_4$ cubes (*cf.* model I in Fig. 17) linked through the *Mo* atoms via one sulfide and two mercaptide,[56a] three mercaptide,[56b−e] or *Fe* (*SR*)$_6$[56f] bridges, on one hand and trinuclear *Mo*−*Fe*−*S* structures (*cf* model II in Fig. 17) such as $[Cl_2FeS_2MoS_2FeCl_2]^{2-}$ [56g] on the other. A different interpretation,[57a] however, led to a yet unknown cluster model shown in Fig. 18a. This model, in which the *Mo* bridges two *Fe*$_4$*S*$_4$ cubes via four sulfur ligands, is consistent with the EXAFS data as well as other spectroscopic evidence.[57a] The *Mo*−*S* distances of 2.35 Å observed in nitrogenase is more consistent with a pseudo-octahedral rather than a tetrahedral coordination where *Mo*−*S* bonds of 2.20-2.25 Å are commonly observed.[56g,57b] Knowing now that the trinuclear *Fe*−*S* cluster *Fe*$_3$*S*$_3$(*SR*)$_6$ with three bridging sulfido ligands is present in some iron-sulfur proteins,[57c] it is natural to consider the cluster models (b) in Fig. 18 as the active site of nitrogenase. This new structural model, in which the *Mo* bridges two *Fe*$_3$*S*$_3$ units, possesses most of the characteristics of model (a). Strong interactions between the *Mo* and the two *Fe*^{A} atoms result in the presumably diamagnetic *Fe*^{A}*S*$_2$*MoS*$_2$*Fe*^{A} unit which can give rise to simple quadrupole doublets in the Mössbauer spectra. The spins (total $S = 3/2$) reside primarily on the *Fe*^{B} atoms. Model (b) has the distinct feature of a variable-size cage through which N_2 can σ-bond to the *Mo* and π-bond to the four high-spin *Fe*^{B} atoms. The net result would be a significant weakening and activation of the *N*−*N* bond. Injection of electrons in a stepwise manner via the two *Fe*$_3$*S*$_3$ units and successive protonation of the terminal nitrogen atom would give rise to the intermediates $Mo-N{\equiv}N \rightarrow Mo-N{=}NH \rightarrow Mo-N-NH_2$ which eventually produces ammonia. We note that a different type of "cage" structure has also been suggested by Lu and coworkers in their Fuzhou model II.[57d] While these proposed models are structurally quite distinct, they are all consistent with the EXAFS findings. This points to the danger of over stretching EXAFS information or over interpreting EXAFS data of unknown systems, particularly in light of the large uncertainties often encountered in the determination of coordination numbers.

Recently, structural organization of calcium in biological molecules has also been investigated by EXAFS spectroscopy.[58] An interesting study of phospholipids, for example, as functions of temperature and content of 0.85 M *CaCl*$_2$ solution has been

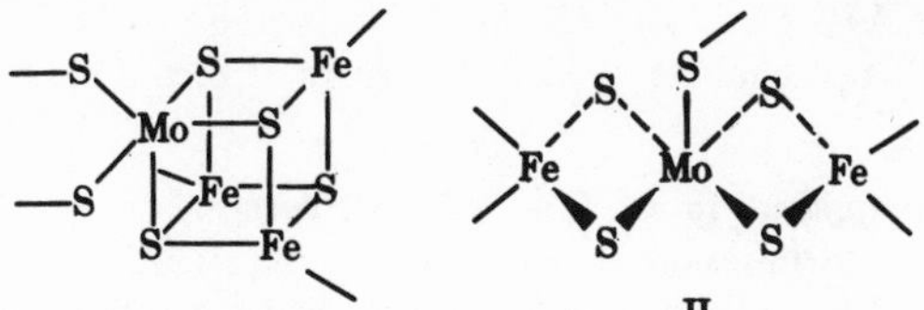

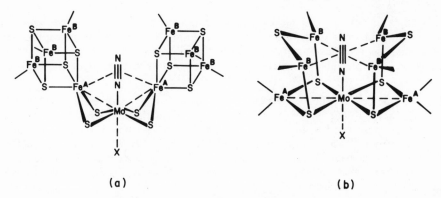

Fig. 17.   Two distinct structural models for the *Mo* site of the nitrogen fixation enzyme nitrogenase proposed by Hodgson and coworkers (from Ref. 55b).

Fig. 18.   Two alternative cluster models for the *Mo* site of nitrogenase: (a) from Ref. 57a and (b) this work. The cysteine sulfurs (or the SR groups) on the irons are omitted for clarity. *X* represents either a sulfur or a lighter atom from a ligand whose nature or existence remains to be determined.

reported. It is shown that for dipalmitoyl phosphatidylcholine, the average $Ca-O$ bond length, ranging from 2.31 (4) to 2.34 (2) Å, remains relatively unchanged for 3%-15% by weight of the $CaCl_2$ solution in the sample, at temperatures below or above the gel transition of the pure lipid water systems. At higher contents of $CaCl_2$ solution (22%-30%) or at higher temperatures (above the gel transition), the observed average $Ca-O$ distance is the same as that observed for the 0.85 M $CaCl_2$ solution which has also been determined by EXAFS to be 2.38 (2) Å. A coordination number of 6 is also suggested. These results imply that at high contents of the $CaCl_2$ solution or at high temperatures, the calcium ions probably exist as the hexaaquacalcium (II) whereas for other solution contents and temperatures, the calcium ions may be coordinated with the oxygen atoms from the phosphate or the carboxylate of the lipid polar groups. Preliminary EXAFS data are also available for calcium-binding proteins such as concanavalin A, thermolysin, muscle calcium-containing parvalbumin (carp), troponin C, and Ca^{2+} ATPase from sarcoplasmic reticulum, etc.[58]

Other published EXAFS studies in biochemistry include those on cytochrome P-450 and chloroperoxidase,[59] cytochrome oxidase,[60] azurin,[61] etc. Further examples can be found in SSRL Reports.[62]

Another important application of EXAFS spectroscopy is to provide information about local structures of drugs before and after their interactions with biological systems. In fact, it may ultimately be possible to study drug actions on living organisms such as intact cells, *in vitro* or *in vivo*. Since EXAFS can focus on the absorbing atoms (usually the heavy atoms), no separation or isolation is required. In this context, we note that several important anti-tumor platinum drugs and their interactions with DNA have been studied by EXAFS.[63]

Inorganic Systems

EXAFS is useful in inorganic chemistry, especially when single crystals are not available or when structural information in solution is sought. It is capable of differentiating between metal-metal and metal-ligand bonds, provided that the atomic numbers of the neighboring metal and ligand atoms are sufficiently different ($\Delta Z \gtrsim 4$).[28]

EXAFS study of the chemically unstable, "one-electron" metal-metal-bonded cobalt cluster $[CpCoPPh_2]_2^+$ ($Cp = \eta^5 - C_5H_5; Ph = C_6H_5$) provided the first structural evidence that the metal-metal bond is significantly weakened upon oxidation of the neutral dimer to the monocation.[64] The neutral dimer (inset of Figure 19) was crystallographically shown to possess a bent Co_2P_2 core of idealized C_{2v} symmetry with an electron-pair $Co-Co$ bond of 2.56(1) Å. Its EXAFS spectrum is shown in Figure 19 (solid curve) along with the three backscattering components ($5C + 2P + 1Co$), which were resolved by curve-fitting. For the monocation, the magnitude of the cobalt backscattering ($1Co$) decreases significantly. Based upon a difference Fourier technique, a small increase in the $Co-Co$ distance of 0.08 Å (from 2.57(1) to 2.65(5) Å) and a significant weakening of the $Co-Co$ bond, as indicated by a significant increase in the Debye-Waller factor (from 0.058(4) to 0.10(2) Å), are observed in going from the neutral parent to the monocation.

Copper oxalate ($Cu^{II} C_2O_4 \cdot 1/3 H_2O$) has been studied by EXAFS in order to determine its structure.[65] Based on the magnetic properties of this compound, which is consistent with a long chain of equivalent and equidistant Cu(II) ions, two structural models have been proposed. The first model has a linear-chain ribbon-like structure similar to that found in $FeC_2O_4 \cdot 2 H_2O$. In the second model, the oxalate ions form bridges between the copper coordination planes (with the chain axis perpendicular to the plane) in a way similar to that found in copper acetate dimer. An important structural difference between these two models is the first metal-metal distance which is greater than 5 Å in the first model and less than 3 Å in the second model. A detailed comparison of the EXAFS of $CuC_2O_4 \cdot 1/3 H_2O$ with that of other structurally known copper compounds with either a long or a short metal-metal distance led to the conclusion that the second shell surrounding the Cu(II) ions is exclusively composed of light atoms. For the first shell, the four $Cu-O$ distances average to 1.98(2) Å. These results are consistent only with the first model.

More recently, EXAFS studies of the polymeric compounds $Cu(C_6O_4X_2)$ and $Cu(C_6O_4X_2)(NH_3)_2$ where $(C_6O_4X_2)^{2-}$ is the chloranilato ($X = Cl$) or the bromanilato ($X = Br$) have been reported.[66] The bromo species are particularly well suited for

EXAFS studies since both the copper and the bromine K edges can be studied. For $Cu(C_6O_4Br_2)$, the copper edge EXAFS provided the distances of 1.95 and 2.67 Å for $Cu-O_1$ and $Cu\cdots C_1$, respectively *(cf.* Ref. 66). The nonbonding $Cu\cdots C_2$ and $Cu\cdots Br$ (5.04 Å) distances were also identified in the Fourier transform. The bromine edge EXAFS, on the other hand, provided the distances of 1.86, 2.84, 3.11, 4.99, and 6.55 Å for $Br-C_2$, $Br\cdots C_1$, $Br\cdots O_1$, $Br\cdots Cu$, and $Br\cdots Br$, respectively. The amplitudes of the $Cu\cdots Br$ and the $Br\cdots Br$ peaks are strongly enhanced by the multiple scattering (focusing) effects due to the linear or nearly linear atomic arrays $Cu-O_1\cdots Br$ and $Br-C_2\cdots C_2'-Br'$, respectively. A detailed comparison of these EXAFS distances with interatomic distances expected for various structural models based on the known geometry of the bromanilato anion led to the conclusion that $Cu(C_6O_4X_2)$ has a planar ribbon-like structure rather than a layer structure. EXAFS studies on $Cu(C_6O_4X_2)(NH_3)_2$ revealed a rhombic distortion of the copper sites upon coordination with two nitrogen-containing ligands such as ammonia giving rise to two short $Cu-N$ distances, two short $Cu-O$ distances, and two long $Cu-O$ distances.

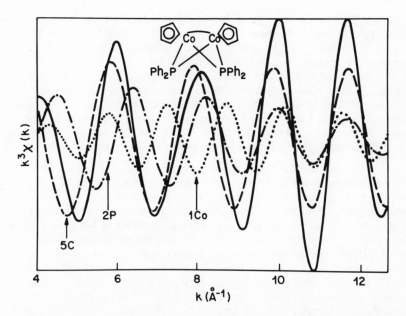

Fig. 19.   A *three-shell* system: the filtered EXAFS spectrum (—) of $[CpCoPPh_2]_2$ is resolved into three backscattering components by curve fitting: $5C$ (---), $2P$ (-.-.-), $1Co$ (...). The fitting model involves a sum of three terms which arise from five $Co-C$, two $Co-P$, and one $Co-Co$ bonds surrounding each cobalt atom. Each term contains the corresponding average distance and the Debye-Waller factor which are least-squares refined. The ordinate scales are -5.93 to 5.33 Å^{-3}.

Other EXAFS applications involving inorganic systems[62,67-75] include iron and molydenum complexes,[30] metallic glasses,[68] superconductors,[69] superionic conductors,[70] semiconductors[71], solutions[72] molten salt[73] and liquid metals.[74] Some of these studies will be described elsewhere in this book.[75]

Another important use of EXAFS is the determination of local structures of the catalytic sites of homogeneous or heterogeneous catalysts (supported or unsupported).[76,77] In a study of the structural variation of the catalytic site of the polymer-bound bromo-Wilkinson's catalyst $(Rh(PPh_3)_3Br)$ as a function of the degree of cross-linking,[76a] the K edges of both rhodium and bromine atoms were measured. For the Rh EXAFS, the characteristic multi-distance interference pattern (with a beat node at $k \approx 9 \text{Å}^{-1}$) required a two-shell, three-distance model $(Rh-P_1, Rh-P_2,$ and $Rh-Br)$ for curve-fitting whereas for the Br EXAFS, a single-shell model $(Br-Rh)$ was adequate. These results suggested a dibromo-bridged dirhodium species $(P_2RhBr_2RhP_2)$ for 2% cross-linking but a monomeric species (RhP_3Br) for 20% crosslinking[76a] of the polymer.

Sinfelt, Lytle, and others[62,77] have studied a large number of supported metal catalysts by EXAFS which have provided a better understanding of their structures. Details of and references to these important studies can be found in a later chapter.[78]

Conclusions

In summary, there are several highly attractive features of the EXAFS technique which makes it a powerful structural tool. First, it is extremely fast. With synchrotron radiation, each spectrum per element takes 10-15 minutes, though for dilute systems, data averaging or specialized instrumentation may be required. With laser-produced X-rays, a spectrum can be recorded on film in a few nanoseconds. Second, both sample preparation and data collection are relatively easy. No single crystals are required. Third, being sensitive to short-range order in atomic arrangements rather than long-range crystalline order, EXAFS can focus on the local environment of specific absorbing atoms. Fourth, EXAFS technique can be used for a wide variety of materials such as amorphous solids, liquids, solutions, gases, polymers, and surfaces.

An obvious deficiency of EXAFS spectroscopy lies in the fact that it does not provide full 3-D structural details. It gives only local structures in terms of 1-D radial distribution functions about each absorber. The absorbing atoms are preferably different in atomic number or equivalent in chemical environment. Any absorbing element appearing in several different chemical states (or environments) will greatly complicate the interpretation of its EXAFS data which is a superposition of all the states. No *direct* method of determining angular information is available (except from single-crystal measurements utilizing polarized X-rays[79]). Furthermore, EXAFS diminishes rapidly beyond the first and second coordination shells (typically $r \gtrsim 4\text{Å}$) except in cases where atoms are nearly colinear. In such cases, EXAFS from atoms as far as 6 Å can be observed due to amplitude enhancement (focusing effect). For these systems, one must therefore take into account multiple scattering processes involving the intervening atom(s). In fact, a new multiple scattering formalism has been developed[23] *(vide supra)* which enables bond angle determinations with an accuracy of 5% or *ca* 5°.

Nevertheless, the structural content of EXAFS is unparalleled by other spectroscopic techniques when one considers that the few most important bonds in a complex system can be probed within minutes. The future of EXAFS spectroscopy is as bright as the future synchrotron radiation sources. Dedicated synchrotron sources with radiation energy ranging from UV to hard X-rays are now available. These highly intense light sources will undoubtedly open up a new era in exciting chemical, biological, and material research.

ACKNOWLEDGEMENTS

The author wishes to thank M. Marcus and A. Simons for many helpful comments.

REFERENCES

1. R. de L. Kronig, Z. Physik., **70,** 317 (1931); **75,** 191, 468 (1932).

2. (a) E. A. Stern, Phys. Rev. B, **10,** 3027 (1974); (b) E. A. Stern, D. E. Sayers, and F. W. Lytle, ibid., **11,** 4836 (1975), and references cited therein.

3. (a) B. M. Kincaid and P. Eisenberger, Phys. Rev. Lett., **34,** 1361 (1975); (b) H. Winick and A. Bienenstock, Ann. Rev. Nucl. Part. Sci., **28,** 33 (1978); (c) I. Lindau and H. Winick, J. Vac. Sci. Tecnol., **15,** 977 (1978); (d) R. E. Watson and M. L. Perlman, Science, **199,** 1295 (1978); (e) B. W. Batterman and N. W. Ashcroft, Science, **206,** 157 (1979).

4. The physics and biological aspects of EXAFS have been reviewed elsewhere by others, see, e.g., (a) E. A. Stern, Contemp. Phys., **19,** 289 (1978); (b) P. Eisenberger and B. M. Kincaid, Science, **200,** 1441 (1978); (c) R. G. Shulman, P. Eisenberger, and B. M. Kincaid, Ann. Rev. Biophys. Bioeng., **7,** 559 (1978); (d) D. R. Sandstrom and F. W. Lytle, Ann. Rev. Phys. Chem., **30,** 215 (1979); (e) S. P. Cramer and K. O. Hodgson, Prog. Inorg. Chem., **25,** 1 (1979); (f) T. M. Hayes, J. Non-Cryst. Solids, **31,** 57 (1978); (g) J. Wong in "Metallic Glasses", H. J. Guntherodt, Ed., Springer-Verlag, Berlin (1980); (h) "Synchrotron Radiation Research", ed. H. Winick and S. Doniach, Plenum, N.Y. (1980); (i) P. A. Lee, P. H. Citrin, P. Eisenberger, and B. M. Kincaid, Rev. Mod. Phys., in press.

5. U. C. Srivastava and H. L. Nigam, Coord. Chem. Rev., **9,** 275 (1972-73).

6. (a) R. G. Shulman, Y. Yafet, P. Eisenberger, & W. E. Blumberg, Proc. Natl. Acad. Sci. U.S.A, **73,** 1384 (1976); (b) F. W. Lytle, P. S. P. Wei, R. B. Greegor, G. H. Via, and J. H. Sinfelt, J. Chem. Phys., **70,** 4849 (1979); (c) L. Powers, W. E. Blumberg, B. Chance, C. H. Barlow, J. S. Leigh, Jr., J. Smith, T. Yonetani, S. Vik, and J. Peisach, Biochim. Biophy. Acta, **546,** 520 (1979).

7. (a) J. Jaklevic, J. A. Kirby, M. P. Klein, A. S. Robertson, G. S. Brown, and P. Eisenberger, Solid State Comm., **23,** 679 (1977); (b) F. S. Goulding, J. M. Jaklevic, and A. C. Thompson, SSRL Report No. 78/04, May 1978; (c) E. A. Stern and S. M. Heald, Rev. Sci. Instrum., **50,** 1579 (1979); (d) P. H. Citrin, P. Eisenberger, and R. Hewitt, Phys. Rev. Lett., **41,** 309 (1978); (e) J. Stohr, D. Denley, and P. Perfetti, Phys. Rev. B, **18,** 4132 (1978); (f) M. Isaacson, J. Chem. Phys., **56,** 1818 (1972); (g) J. I. Ritsko, S. E. Schnatterly and P. G. Gibbons, Phys. Rev. Lett., **32,** 671 (1974); (h) R. A. Bonham in "Momentum Wave Functions-1976", American Institute of Physics, New York (1977); (i) B. M. Kincaid, A. E. Meixner, and P. M. Platzman, Phys. Rev. Lett., **40,** 1296 (1978); (j) D. C. Joy and D. M. Maher, Science **206,** 162 (1979).

8. (a) C. A. Ashley and S. Doniach, Phys. Rev. B, **11**, 1279 (1975). (b) P. A. Lee and G. Beni, Phys. Rev. B, **15**, 2862 (1977); (c) P. A. Lee and J. B. Pendry, ibid. **11**, 2795 (1975).

9. Though simply defined as $\chi = (\mu - \mu_o)/\mu_o$ (where μ and μ_o are the observed and 'free atom' absorption coefficients, respectively), the determination of χ (generally termed as 'background substration') is by no means straightforward since μ_o is generally not known. A general procedure is to approximate μ_o by a smooth curve (some polynomial or spline) fitted to μ. In transmission experiments, μ generally drops off monotonically due primarily to the energy dependence of the ionization chamber efficiency and the absorption due to other atoms. In fluorescence experiments, on the other hand, the baseline rises as a function of energy owing to increasing sample penetration, increased Compton scattering, reduced absorption of the scattering, and other effects.

10. It should be cautioned that the Debye-Waller factor as determined by EXAFS is different from that implied by conventional crystallography in that it refers to the root-mean-square *relative* displacement *along the bond direction* and not the absolute root-mean-square displacement of individual atoms. For the first shell, the motions are significantly correlated whereas for higher shells the correlation is greatly reduced.

11. S. J. Cyrin, *Molecular Vibrations and Mean Square Amplitudes,* Elsevier, Amsterdam, 1968, p. 77.

12. G. Beni and P. M. Platzman, Phys. Rev. B, **14**, 1514 (1976).

13. R. B. Greegor and F. W. Lytle, Phys. Rev B, **20**, 4902 (1979).

14. E. Sevillano, H. Meuth, and J. J. Rehr, Phys. Rev. B, **20**, 4908 (1979).

15. P. Eisenberger and G. S. Brown, Solid State Commun., **29**, 481 (1979).

16. T. M. Hayes and J. B. Boyce, Chapter 5 of this book.

17. T. M. Hayes, J. B. Boyce, and J. L. Beeby, J. Phys. C, **11**, 2931 (1978).

18. E. D. Crozier, Chapter 6 of this book.

19. E. D. Crozier and A. J. Seary, Can. J. Phys., **58**, 1388 (1980).

20. J. J. Rehr, E. A. Stern, R. L. Martin, and E. R. Davidson, Phys. Rev. B, **17**, 560 (1978).

21. (a) E. A. Stern, B. Bunker, and S. M. Heald, Chapter 4 of this book; (b) E. A. Stern, B. A. Bunker, and S. M. Heald, Phys. Rev. B, **21**, 5521 (1980).

22. P. Eisenberger and B. Lengeler, to be published in Phys. Rev. B (1980).

23. B. K. Teo, J. Am. Chem. Soc., submitted for publication.

24. C. J. Powell, Surface Sci., **44**, 29 (1974).

25. D. R. Penn, Phys. Rev. B, **13**, 5248 (1976).

26. M. P. Seah and W. A. Dench, Surface Inter. Anal., **1**, 2 (1979).

27. Strictly speaking, the single-electron single-scattering theory of EXAFS already includes one particular multiple scattering correction: *viz.,* the backscattering process involving the central atom which gives rise to the $2kr$ phase factor. In this paper, "multiple scattering" refers to processes involving atoms other than the central atom.

28. B. K. Teo and P. A. Lee, J. Am. Chem. Soc., **101**, 2815 (1979).

29. (a) P. H. Citrin, P. Eisenberger, and B. M. Kincaid, Phys. Rev. Lett., **36**, 1346 (1976); (b) B. K. Teo, P. A. Lee, A. L. Simons, P. Eisenberger, and B. M. Kincaid, J. Am. Chem. Soc., **99**, 3854 (1977); (c) P. A. Lee, B. K. Teo, and A. L. Simons, ibid, **99**, 3856 (1977).

30. S. P. Cramer, T. K. Eccles, F. Kutzler, K. O. Hodgson, and S. Doniach, J. Am. Chem. Soc., **98**, 8059 (1976).

31. R. G. Shulman, P. Eisenberger, W. E. Blumberg, N. A. Stombaugh, Proc. Nat. Acad. Sci. U.S.A., **72**, 4002 (1975).

32. G. Martens, P. Rabe, N. Schwentner, and A. Werner, Phys. Rev. Lett., **39**, 1411 (1977).

33. P. A. Lee, Chapter 2 of this book.

34. A. Bienenstock, Chapter 14 of this book.

35. B. W. Batterman, Chapter 15 of this book.

36. J. B. Hastings, Chapter 16 of this book.

37. P. J. Mallozzi, R. E. Schwerzel, H. M. Epstein, and B. E. Campbell, Science, **206**, 353 (1979).

38. G. S. Knapp, H. Chen, T. E. Klippert, Rev. Sci. Instrum., **49**, 1658 (1978).

39. For more detailed discussions on S/N of various EXAFS techniques, see Ref. 4i.

40. J. B. Hastings, P. Eisenberger, B. Lengeler, and M. L. Perlman, Phys. Rev. Lett., **43**, 1807 (1979).

41. M. Marcus, L. S. Powers, A. R. Storm, B. M. Kincaid, and B. Chance, Rev. Sci. Instrum., **51**, 1023 (1980).

42. See Chapters 17-21 of this book.

43. R. D. Leapman and V. E. Cosslet, J. Phys. D, **9**, 25 (1976).

44. P. E. Batson and A. J. Craven, Phys. Rev. Lett., **42**, 893 (1979).

45. For an excellent review, see R. H. Holm, Acc. Chem. Res., **10**, 427 (1977).

46. R. G. Shulman, P. Eisenberger, B. K. Teo, B. M. Kincaid, and G. S. Brown, J. Mol. Biol., **124**, 305 (1978), and references cited therein.

47. B. K. Teo, R. G. Shulman, G. S. Brown, and A. E. Meixner, J. Am. Chem. Soc., **101**, 5624 (1979).

48. K. D. Watenbaugh, L. C. Sieker, J. R. Herriot, and L. H. Jensen, Acta Crystallogr. B, **29**, 943 (1973).

49. (a) B. Bunker and E. A. Stern, Biophys. J., **19**, 253 (1977); (b) D. E. Sayers, E. A. Stern, and J. R. Herriott, J. Chem. Phys., **64**, 427 (1976).

50. (a) P. Eisenberger, R. G. Shulman, G. S. Brown, and S. Ogawa, Proc. Natl. Acad. Sci., U.S.A., **73**, 491 (1976); (b) P. Eisenberger, R. G. Shulman, B. M. Kincaid, G. S. Brown, and S. Ogawa, Nature (London), **274**, 30 (1978).

51. (a) J. L. Hoard, Science, **174**, 1295 (1971); (b) M. F. Perutz, Nature (London), **228**, 726 (1970).

52. G. S. Brown, G. Navon, and R. G. Shulman, Proc. Natl. Acad. Sci., U.S.A., **74**, 1794 (1977).

53. S. M. Heald, E. A. Stern, B. Bunker, E. M. Holt, and S. L. Holt, J. Am. Chem. Soc., **101**, 67 (1979).

54. J. M. Brown, L. Powers, B. Kincaid, J. A. Larrabee, and T. G. Spiro, J. Am. Chem. Soc. **102**, 4210 (1980).

55. (a) T. E. Wolff, J. M. Berg, C. Warrick, K. O. Hodgson, R. H. Holm, and R. B. Frankel, J. Am. Chem. Soc., **100**, 4630 (1978); (b) S. P. Cramer, K. O. Hodgson, W. O. Gillum, and L. E. Mortenson, *ibid.*, **100**, 3398 (1978); (c) S. P. Cramer, W. O. Gillum, K. O. Hodgson, L. E. Mortenson, E. I. Stiefel, J. R. Chisnell, W. J. Brill, and V. K. Shah, *ibid.*, **100**, 3814 (1978).

56. (a) T. E. Wolff, J. M. Berg, C. Warrick, K. O. Hodgson, and R. H. Holm, J. Am. C. S., **100**, 4630 (1978); (b) T. E. Wolff, J. M. Berg, K. O. Hodgson, R. B. Frankel, and R. H. Holm, *ibid.*, **101**, 4140 (1979); (c) G. Christou, C. D. Garner, F. E. Mabbs and T. J. King, J. C. S. Chem. Commun., 740 (1978); (d) G. Christou, C. D. Garner, F. E. Mabbs, and M. G. B. Drew, *ibid.*, **91** (1979); (e) S. R. Acott, G. Christou, C. D. Garner, T. J. King, F. E. Mabbs, and R. M. Miller, Inorg. Chim. Acta., **35**, L337 (1979); (f) T. E. Wolff, J. M. Berg, P. P. Power, K. O. Hodgson, R. H. Holm, and R. B. Frankel, *ibid.*, **101**, 5454 (1979); (g) D. Coucouvanis, N. C. Baenziger, E. D. Simhon, P. Stremple, D. Swenson, A. Simopoulos, A. Kostikas, V. Petrouleas, and V. Papaefthymiou, J. Am. Chem. Soc., **102**, 1732 (1980).

57. (a) B. K. Teo and B. A. Averill, Biochem. Biophys. Res. Commun., **88**, 1454 (1979); (b) R. H. Tieckelmann, H. C. Silvis, T. A. Kent, B. H. Huynh, J. V. Waszczak, B. K. Teo, and B. A. Averill, J. Am. Chem. Soc., **102**, 5550 (1980); (c) C. D. Stout, D. Ghosh, V. Pattabhi, and A. Robbins, J. Biol. Chem., **255**, 1797 (1980); (d) Fujian Institute of Research on the Structure of Matter (PRC), private communication.

58. L. Powers, P. Eisenberger, and J. Stamatoff, Ann. N.Y. Acad. Sci., **307**, 113 (1978).

59. S. P. Cramer, J. H. Dawson, K. O. Hodgson, and L. P. Hager, J. Am. Chem. Soc., **100**, 7282 (1978).

60. V. W. Hu, S. I. Chan, and G. S. Brown, Proc. Natl. Acad. Sci. U.S.A., **74**, 3821 (1977).

61. T. Tullius, P. Frank, and K. O. Hodgson, Proc. Natl. Acad. Sci. U.S.A., **75**, 4069 (1978).

62. For example, SSRL Publication List; SSRL Reports 78/10, 79/03, 79/10, 80/01, etc.

63. (a) B. K. Teo, K. Kijima, and R. Bau. J. Am. Chem. Soc., **100,** 621 (1978), and references cited therein. (b) B. K. Teo, P. Eisenberger, J. Reed, J. K. Barton, S. J. Lippard, J. Am. Chem. Soc., **100,** 3225 (1978), and references cited therein.

64. B. K. Teo, P. Eisenberger, and B. M. Kincaid, J. Am. Chem. Soc., **100,** 1735 (1978).

65. A. Michalowicz, J. J. Girerd, and J. Goulon, Inorg. Chem., **18,** 3004 (1979).

66. M. Verdaguer, A. Michalowicz, J. J. Girerd, N. Alberding, and O. Kahn, Inorg. Chem., **19,** 3271 (1980).

67. G. Martens, P. Rabe, N. Schwentner, and A. Werner, Phys. Rev. B, **17,** 1481 (1978).

68. (a) H. S. Chen, B. K. Teo, and R. Wang, Abst. 4th Inter. Conf. Liq. Amor. Metals, Grenoble, France, July 7-11, 1980; (b) J. Wong, F. W. Lytle, R. B. Greegor, H. H. Liebermann, J. L. Walter, and F. E. Luborsky, Proc. 3rd Inter. Conf. Rapid. Quench. Metals, Sessex University, Vol. II, 1978, p. 345; (c) T. M. Hayes, J. W. Allen, J. Tauc, B. C. Giessen, and J. J. Hauser, Phys. Rev. Lett., **40,** 1282 (1978); (d) E. A. Stern, S. Rinaldi, E. Callen, S. Heald, and B. Bunker, J. Mag. Mater., **7,** 188 (1978).

69. (a) G. S. Brown, L. R. Testardi, J. H. Wernick, A. B. Hallak, and T. H. Geballe, Solid State Commun., **23,** 875 (1977).

70. J. B. Boyce and T. M. Hayes in Chapter 2 of *Physics of Superionic Conductors,* ed. M. B. Salamon, Vol. 15 of Topics in Current Physics, Springer-Verlag, Berlin (1979).

71. (a) S. H. Hunter, A. Bienenstock, and T. M. Hayes, in *The Structure of Non-Crystalline Materials,* ed. P. H. Gaskell, Taylor and Francis, London, 1977, p. 73; (b) S. H. Hunter, A. Bienenstock, and T. M. Hayes, in *Amorphous and Liquid Semiconductors,* ed. W. E. Spear, Univ. Edinburgh, Edinburgh, 1977, p. 78.

72. (a) P. Eisenberger and B. M. Kincaid, Chem. Phys. Lett., **36,** 134 (1975); (b) D. R. Sandstrom, H. W. Dodgen and F. W. Lytle, J. Chem. Phys., **67,** 473 (1977); (c) D. R. Sandstrom, J. Chem. Phys., **71,** 2381 (1979).

73. (a) E. D. Grozier, F. W. Lytle, D. E. Sayers, and E. A. Stern, Can J. Chem., **55,** 1968 (1977); (b) J. Wong and F. W. Lytle, J. Non-Cryst. Solid., **37,** 273 (1980).

74. E. D. Crozier and A. J. Seary, Can. J. Phys., **58,** 1388 (1980).

75. See Chapters 7 to 12 of this book.

76. (a) J. Reed, P. Eisenberger, B. K. Teo, and B. M. Kincaid, J. Am. Chem. Soc., **100,** 2375 (1978); (b) J. Reed, P. Eisenberger, B. K. Teo, and B. M. Kincaid, ibid., **99,** 5217 (1977).

77. (a) J. H. Sinfelt, G. H. Via, and F. W. Lytle, J. Chem. Phys., **68,** 2009 (1978); (b) F. W. Lytle, NBS Spec. Publ. U.S.A., **475,** 34 (1977); (c) I. Bassi, F. W. Lytle, and G. Parravano, J. Catal., **42,** 139 (1976); (d) G. H. Via, J. H. Sinfelt, F. W. Lytle, J. Chem. Phys., **71,** 690 (1979); (e) J. H. Sinfelt, G. H. Via, and F. W. Lytle, J. Chem. Phys., **72,** 4832 (1980); (f) Ref 6c.

78. G. H. Via, J. H. Sinfelt, and F. W. Lytle, Chapter 10 of this book.

79. (a) S. M. Heald and E. A. Stern, Phys. Rev. B, *16*, 5549 (1977); (b) L. I. Johansson and J. Stohr, Phys. Rev. Lett., **43**, 1882 (1979).

UNDERSTANDING THE CAUSES OF

NON-TRANSFERABILITY OF EXAFS AMPLITUDE

E. A. Stern, B. Bunker and S. M. Heald[†]

Physics Dept. FM-15
University of Washington
Seattle, WA 98195

I. Introduction

The use of the extended x-ray absorption fine structure (EXAFS) technique has expanded rapidly in recent years.[1,2] EXAFS provides information about the local environment of the x-ray absorbing atom. In addition to the interatomic distances r, the mean square deviation, σ^2, of r and the number of atoms, N, at r can be extracted from an EXAFS spectrum. However, in order to extract N and σ^2 the amplitude of the EXAFS spectrum must be calibrated. This has been done using theoretical calculations[3] or empirically using known materials as standards,[4] but depends critically on the accuracy of transferability from sample to sample. It is the aim of this paper to assess experimentally the accuracy of amplitude transferability and to understand some of the factors that may cause variability.

The K edge EXAFS spectrum from an unoriented sample can be represented in the following form:[5,6]

$$\chi(k) = \frac{\mu - \mu_0}{\mu_0} = \sum_i \frac{N_i F_i(k)}{k \, r_i^2} \, P_i(k) e^{-2r_i/\lambda} \, sin(2kr_i + \phi_i(k)) \tag{1}$$

where μ is the x-ray absorption coefficient, μ_0 the smoothly varying background absorption, k is the wavevector of the outgoing photoelectron, λ is the electron mean free path, and the sum is over the coordination shells of neighbors to the absorbing atom. $P_i(k)$ is the modulation of $\chi(k)$ induced by the pair distribution function of the atoms in the ith shell. In the case that the atoms are distributed in a symmetric manner $F(k)$ and $\phi(k)$ are the amplitude and phase functions as defined by Teo and Lee.[3a] Assymetric distributions add another contribution to $\phi(k)$. Equation (1) is based on a one electron picture, and has a relatively simple form because most of the complexities of the process are hidden in the functions $F(k)$ and $\phi(k)$. To extract N and σ^2 it is $F(k)$

[†] Present address, Brookhaven National Lab., Upton, N.Y. 11973.

which must be first determined theoretically or experimentally.

The theoretical calculations have been quite successful in reproducing the qualitative behavior of EXAFS amplitudes, but quantitatively they generally overestimate the overall amplitude. This is not unexpected since many electron effects are left out of the calculations, and these would be expected to reduce the EXAFS amplitudes. However, in using the calculations it is usually assumed that the reduction factor is a constant independent of k. A careful study of the EXAFS from Br_2 molecules[7] has shown that the reduction factor is definitely not a constant in that case, casting doubt on the reliability of the assumption in other cases.

The experimental determination of $F(k)$ relies on its transferability from a known system to an unknown. This seems to be the case for systems which are chemically similar. However, as the applications of EXAFS expand it is sometimes difficult to find similar standard systems. Therefore, the present work was undertaken to better define the accuracy of transferability. This question is also important for the use of the calculations since typically they are done for only a particular chemical state.

II. EXAFS Amplitudes

There are two physically distinct mechanisms of amplitude reduction induced by many electron effects. One, as isolated in Br_2, is a many body effect associated with the absorbing atom. The dipole matrix element for the absorption of a photon includes, in addition to the electron directly acted upon by the dipole operator, another factor of the overlap of the initial and final wavefunctions of the rest of the atomic electrons, the so-called passive electrons. Since the final state of the atom has a hole induced by the dipole transition of the active electron, the passive electrons sense a different potential than in the initial state and their wave function is correspondingly modified. In the energy range above several hundred eV past the absorption edge threshold, the sudden approximation is valid for valence and shallow core states. The contribution of the passive electrons is to decrease the one electron matrix element since the overlap of their final state wave function with their initial state is less than one.

The amount of the reduction from this first mechanism can be written as[8,3b]

$$S_0^2 \equiv |<\phi'_{N-1}|\phi_{N-1}>|^2 \tag{2}$$

where ϕ'_{N-1} and ϕ_{N-1} are the N-1 electron wave functions describing the passive electrons before and after the formation of the core hole. The factor S_0^2 has been calculated for many atoms and is found to lie in the range 0.7 to 0.8 for greater than about 200 eV excitation energies.[9]

The second mechanism of amplitude reduction is the finite lifetime of the excited state. This lifetime is a convolution of the lifetime of the core hole left behind in the atom and the lifetime of the photoelectron. Since EXAFS is caused by the interference of the backscattered portion of the photoelectron's wave function with its outgoing part, the excited state must last long enough to permit the photoelectron to travel to the scattering atom and back. If the state decays before this transit time the wave function loses coherence and no interference can occur. The mean free path term λ in Eq. (1) accounts for this lifetime effect and it is expected, in general, that λ is $\lambda(k)$.

These two many body mechanisms modify the one electron expression for $\chi(k)$ in distinct ways. The first effect modifies $F(k)$ from its one electron form as shown directly by measurements on Br_2. The second effect determines $\lambda(k)$ and is distinguished from the first effect by adding r_i-variation to $\chi(k)$ in addition to a k-dependence.

To date, the most comprehensive EXAFS calculations have been carried out by Lee and coworkers.[3] They use a one-electron approach, and a local scattering approximation based on the Kohn-Sham density functional formalism to treat the backscattering atom. They attempt to include inelastic processes in the backscattering atom by constructing a complex effective potential to treat the scattering. While this provides an approximate treatment of the inelastic processes for the backscattering atom, it does not include effects which originate within the central atom. It also does not include the new channels of inelastic processes opened by placing atoms in the condensed state: because the excitation gap is generally lowered in condensed matter systems, new low energy excitations are made possible. As an example, in the metallic state the electron gas scattering contributes strongly to λ.

Since there is the dipole sum rule which states that the total absorption must remain the same independent of multi-electron effects, other absorption processes must compensate for the loss of intensity in the primary channel caused by the first mechanism. These are the so-called "shake-up" and "shake-off" or correlation processes in which one or more passive electrons are excited along with the photoelectron.[9] From conservation of energy, this implies that the photoelectron has less energy. It can still have EXAFS associated with it, but the EXAFS is shifted in energy and may have a different phase shift associated with it. It was shown[3b,8] that this results in a washing out of these channels, and typically there is relatively little contribution to the total EXAFS spectrum from these multi-electron excitations. The exception to this conclusion is the case where the system contains a significant fraction of low energy ($\lesssim 10$ eV) excitations. Then these channels can contribute significantly to the EXAFS spectrum, especially at high energies where the small energy shift is insignificant.

These ideas are beautifully demonstrated in the EXAFS from Br_2 molecules.[7] In this case the multi-electron excitation spectrum has two components; an atomic like contribution consisting mainly of shake-off processes, and excitations of molecular states which are mainly low energy. At high energies it was found that the EXAFS amplitudes agreed with the theory of Lee reduced by a factor S_0^2 appropriate to a Br *atom*. Because the molecular excitations are low energy they do not cause a significant reduction of the EXAFS even though calculations show they make a substantial contribution to S_0^2. The Br_2 measurements also demonstrate another important point. The above ideas are based on the sudden approximation which is appropriate at high energies. At low energies the adiabatic approximation which predicts no reduction is more appropriate. Indeed for Br_2 the EXAFS amplitudes agree with the theory of Lee at low energies and drop smoothly to the high energy result. Other measurements[10] have also shown that the multi-electron excitations do not turn on until an energy of several times their threshold energy is reached.

One important question to answer is how sensitive the overlap effect, and thus $F(k)$, is to changes in the local chemical environment. It is known from x-ray photoemission measurements that the multi-electron excitation spectrum[11] is sensitive to variations in chemical environment and thus S_0^2 could also vary significantly. Are the same considerations used in explaining the Br_2 results also valid in other cases, which would then cancel the chemical effects and make $F(k)$ transferable?

To help answer this question we have studied the isostructural compounds $CoCl_2$, $FeCl_2$, and $MnCl_2$. Each is observed by x-ray photoemission measurements[11] to have significant low energy correlation state satellites in addition to the primary channel. As shown in Table I the strength of these satellites is quite different while the atomic S_0^2 is nearly the same for Co, Fe, and Mn. Thus, these materials provide a good test of the importance of changes in the low energy correlation state excitations induced by chemical variations in affecting $F(k)$. We have found that the Cl backscattering is the same for all these three samples indicating that the low energy satellites do not appreciably affect the EXAFS.

We now turn to the second class of factors affecting the EXAFS amplitude, the mean free path effects. The distinguishing feature of these mean free path effects is that they increase in magnitude as r is increased. To investigate this point we have also studied the series of materials comprising the diamond structure Ge and the zincblende structures $GaAs$, $ZnSe$, and $CuBr$. These materials have a very similar structure, with the two Ge atoms in the unit cell of the diamond structure being replaced by different atoms in the zincblende structure without much change in interatomic distances. The variation that occurs in this series is thus mainly limited to increasing the ionicity of the bond with the attendant increase in energy gap between the valence and conduction bands. As far as EXAFS is concerned, the various atoms in the series are quite similar, introducing only small changes in the parameters that enter Eq. (1). From a study of this series we will be able to show that many electron effects produce a significant modification to the independent particle model expression of Eq. (1) and that λ varies considerably in the series.

III. Experimental

The measurements to be discussed in this paper were made on the K-edges of the transition metals in $MnCl_2$, $FeCl_2$, and $CoCl_2$, on the K-edges of both of the constituent atoms of $CuBr$, $ZnSe$, and $GaAs$ and on the K-edges of pure Ge.

Since several experimental problems can affect the measured EXAFS amplitude, it is important to distinguish real amplitude differences from spurious experimental effects. In this section the various problems are described along with the checks which were made to be sure that they are not affecting the results which are presented.

TABLE I Relative energies and intensities of satellite peaks from photoemission experiments (from reference 11). ΔE is the shift of the satellite from the elastic peak, and its intensity relative to the elastic peak is denoted in the second column. S_0^2 is the atomic overlap factor for the transition metal atom.

	ΔE (eV)	$I(\text{sat})/I(\text{main})$	S_0^2
$MnCl_2$	5.1	0.38	0.68
$FeCl_2$	4.7	0.48	0.69
$CoCl_2$	5.1	0.70	0.70

The most fundamental check to make is that of the sample itself. It is important that during sample preparation and measurement the chemical state of the sample is preserved. For the tetrahedral semiconductors this is not a problem since they are stable chemically. However, it was a problem for the dichlorides. The dichlorides are hydroscopic and care had to be taken to insure the sample integrity. These samples were handled only under dry nitrogen. They were ground up, mixed with vacuum grease or "5-minute" epoxy in the proper proportion, and sealed in copper holders using an indium seal. The cells had kapton x-ray windows sealed with epoxy. This arrangement allowed low temperature to room temperature measurements to be made without fear of exposure to the ambient atmosphere.

It was also necessary to check the state of the samples as supplied. After careful cross check of amplitude versus exposure to the atmosphere it was possible to isolate the data corresponding to the good samples.

Other factors which can affect the EXAFS amplitudes are preferred orientation and inhomogeneity of the sample, and contamination of the x-ray beam with harmonics. For all of the materials in this work their symmetry is sufficient to insure no orientation dependence for the first shell EXAFS. For the dichlorides there could be some orientational effects for higher shells, but these shells were not analyzed.

The chief difficulty from inhomogeneity comes if cracks or pinholes exist in the sample which allow part of the x-ray beam to pass through unattenuated. To check the dichlorides several samples were measured for each compound, and the samples were rotated. This rotation should change the apparent inhomogeneities presented to the x-ray beam. No amplitude changes were observed indicating sample inhomogeneities were not a problem for the dichlorides.

The tetrahedral semiconductors samples were made in several ways. All the methods begin by grinding the samples to a fine powder, which is then sieved through a 375 or 400 mesh. In the first method the samples are then dusted uniformly onto 0.0013 cm thick kapton tape. Several layers of tape were then used to make a complete sample. For samples to be measured at elevated temperatures, the powdered samples were mixed with boron nitride powder and packed into boron nitride sample holders. Again, various samples were measured with no significant amplitude changes being observed. Other methods included mixing the powders with vacuum grease or epoxy.

The presence of harmonics in the x-ray beam is a potentially serious problem since they pass through the sample almost unattenuated. In this respect the presence of harmonics is similar to having cracks or pinholes in the sample. The best way to check for the effects of harmonics is to vary the x-ray thickness of the sample, since the effects become small for thin samples. This was done for the dichlorides and tetrahedral semiconductor samples.

The measurements were all made at the Stanford Synchrotron Radiation Laboratory (SSRL) on EXAFS Lines I and II. On Line I a Si 220 channel cut monochromator was used and on Line II a double crystal monochromator with either Si 111 or Si 400 reflection was used. The incident and final x-ray intensities were measured using ionization chambers filled with appropriate gases. Additional checks which can be made are: varying the gas mixture in the rear ionization chamber which changes the detection efficiency for harmonics, and changing the monochromator transmission function by changing the monochromator crystals or relative orientation of the crystals in the case of the two-crystal monochromator. These checks were made for both the dichlorides and tetrahedral semiconductors. Only those measurements for which it is obvious that harmonics are not a problem are used in this paper.

As is obvious from the preceding discussion, a large number of measurements was made. Some examples of the EXAFS $\chi(k)$ obtained are shown in Figs. 1,2 for the materials measured.

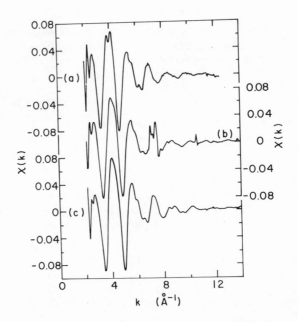

Fig. 1. $\chi(k)$ for (a) $MnCl_2$, (b) $FeCl_2$, and (c) $CoCl_2$, all at 80K.

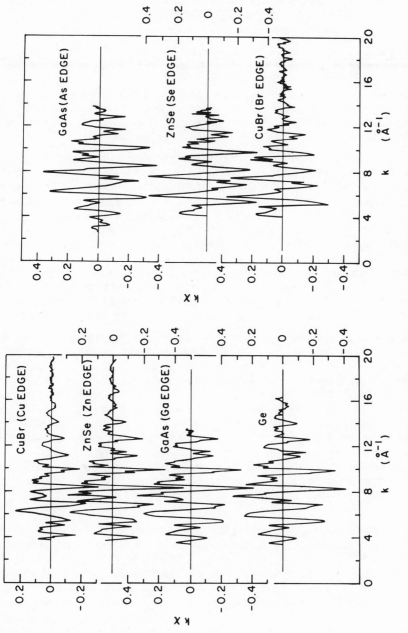

Fig. 2. $\chi(k)$ weighted with k for the tetrahedral semiconductors. The *CuBr* and *ZnSe* data were measured at 10K, the *GaAs* and *Ge* were measured at 80K.

IV. Data Analysis

The extraction of $\chi(k)$ as shown in Figs. 1,2 was done using standard techniques of background subtraction and normalization.[4] The fits of McMaster[12] were used to account for the fall off of μ_0 past the edge. This is necessary when comparing the experimental $\chi(k)$ with theory and is an approximately 10-20% correction at high k.

The next step in the data analysis is to calculate Fourier transforms of the data. In order to simplify the analysis the EXAFS contributions from the individual shells is isolated by Fourier filtering. The peak corresponding to the shell of interest is back transformed into k-space where its amplitude and phase can be determined as a function of k. A typical r-space window is shown for the first and second shells in Fig. 3. A Hanning function is used to provide a smooth termination of the window. Typically it is applied to the outer $0.2 - 0.4$Å on both sides of the window.

Once a given shell is isolated it is easy to extract the k-dependent amplitude of the $\chi(k)$ and correct it for N and r^2. However, two important corrections remain before $F(k)$ can be obtained. The first is to correct for the inevitable distortion of the amplitude caused by the finite filtering range used. This is done by comparing amplitudes which have been obtained using the same r-space and k-space ranges in the above procedure. For example, in comparing the experimental amplitude with theory, the theory is used to generate a $\chi(k)$ which is then analyzed in exactly the same way. The ratio of the analyzed theory with the original can be used to determine the distortion or, more simply, the experimental amplitude can be compared directly with the analyzed theoretical amplitude. To check the success of this procedure the k and r-space ranges can be varied to change the distortion, and the resulting ratios compared. This was done for all cases in this paper, and the resulting differences are included in the error estimates.

The second major correction which must be applied is to account for the disorder term. For all of the materials in this paper the disorder can be accounted for by a Debye-Waller like factor $e^{-2k^2\sigma^2}$. Even at low temperatures σ^2, which is essentially entirely zero point motion, must still be accounted for when comparing different materials. Since the EXAFS σ^2 is different from u^2 normally measured in diffraction experiments, it is not generally known and represents the major complication in determining $F(k)$. For simple molecules σ^2 can be accurately calculated from the measured vibrational properties of the molecule. For the simple solid Ge, σ^2 can be calculated using force constant modes for the lattice dynamics.[13] The accuracy of these calculations for the zero point component, σ_o^2, can be estimated by comparing the temperature dependence of σ^2 with the calculated dependence. For the other compounds in this paper the temperature variation of σ^2 can be used to estimate σ_0^2 using an Einstein or Debye model.

For solids with more than one atom in the unit cell, the contribution to EXAFS σ^2 of the first shell is dominated by high frequency optical modes which should be well represented by an Einstein model. This has been demonstrated for Ge.[13] In fact, the Einstein model has been shown to be an excellent approximation, even for the monatomic solid Cu,[14] and, thus, it should work well for the first shell of the diatomic dichloride compounds. For second and higher shell data, the longer wavelength acoustic modes are more important. Because of this, a Debye model gives a significantly better fit to the observed temperature dependence and has therefore been used to estimate the second-shell zero-point motion for the tetrahedral semiconductors.

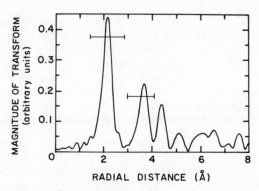

Fig. 3.   Magnitude of the Fourier transform of $k\chi(k)$ for *Ge* measured at 80K.
Also shown are typical ranges for backtransforming first- and second-shell
data.

We now consider the detailed results obtained for each class of materials.

A. Dichlorides

Details of the analysis are given in Ref. 15. The conclusion is that chemical effects
on S_0^2 for this series is negligible and the *atomic* value can be used.

B. Tetrahedral Semiconductors

The diamond and zincblende structure series is a relatively simpler system to
analyze than the dichlorides. Each atomic site is surrounded by a tetrahedral arrangement of atoms. As mentioned above, the main variation through the series is the
degree of ionicity. In Phillips scale,[16] the ionicity of the materials vary from 0.0 for *Ge*
to 0.735 for *CuBr*. The ionicities, lattice constants, and band gaps are listed in Table II.
The low temperature $\chi(k)$ data are shown in Figs. 2a and 2b.

TABLE II - Properties of tetrahedral Semiconductors

Compound	Ionicity[a]	Lattice constant (Å)	Band gap[b] (minimum) (eV)
Ge	0	5.6570	0.80
GaAs	0.310	5.6537	1.35
ZnSe	0.679	5.6676	2.67
CuBr	0.735	5.6905	2.99

a. Reference 16

b. D. L. Greenaway and G. Harbeke, *Optical Properties and Band Structures of Semiconductors,* Pergamon, N.Y., N.Y. (1968).

The first and second shell nearest neighbors are well separated in these tetrahedral materials as shown by the Fourier transform of the $\chi(k)$ data for *Ge* in Fig. 3. As is always the situation, no multiple scattering effects are present in the first shell[17] and, in addition, the second shell is not only relatively prominent, multiple scattering effects are negligible there also. This is predicted theoretically[17] and we find no evidence of multiple scattering experimentally. Thus, inverse transforms of these first two shells can be made directly without the complications of eliminating interference from neighboring shells and the results are not marred by multiple scattering contributions.

In order to determine the Debye-Waller factors for these materials, an Einstein model for the vibrations was fitted to the first shell data where the high-frequency optical modes dominate the phonon spectra. For the second shell, where the longer wavelength modes are important, a Debye model was found to give a better fit to the measured temperature dependence. From these fits the zero point contribution can be found. For *Ge* and *GaAs* the fit was made to three temperatures, one at 80°K, another at room temperature, and the third at 180°K. For *ZnSe*, the fit was made to measurements at room temperature, 160°K, 80°K, 60°K, and 10°K. For *CuBr*, measurements at 80°K, 60°K, 30°K, and 10°K were used; higher temperatures were not appropriate for this purpose because of anharmonic effects. The results are presented in Table III. A check can be made in the case of *Ge*, where the calculations of Rehr et al. have

TABLE III. Debye-Waller disorder σ^2 for the tetrahedral semiconductors. The absolute σ^2 is estimated from the measured change between temperatures T_1 and T_2 using the Einstein approximation for first shell and the Debye approximation for second shell.

Material	Edge	Shell	$T_1(°K)$	$T_2(°K)$	$\sigma^2(T_2)-\sigma^2(T_1)(Å^2)$	$\sigma^2_{FIT}(T_1)(Å^2)$
CuBr	Cu,Br	1	10	80	0.00083 ± 0.00015	0.0036 ± 0.0002
ZnSe	Zn,Se	1	80	295	0.0032 ± 0.0001	0.0025 ± 0.0001
GaAs	Ga,As	1	80	295	0.0021 ± 0.0001	0.0022 ± 0.0001
Ge	Ge	1	80	295	0.0016 ± 0.0001	0.00190 ± 0.00005
Ge (theory)	Ge	1	80	295	0.0015 ± 0.0001[a]	0.00183 ± 0.00001[a]
CuBr	Cu	2	10	80	0.0076 ± 0.0020	0.0067 ± 0.0009
CuBr	Br	2	10	80	0.0038 ± 0.0001	0.0051 ± 0.0001
ZnSe	Zn	2	10	295	0.0110 ± 0.0020	0.0050 ± 0.0006
ZnSe	Se	2	80	295	0.0094 ± 0.0028	0.0049 ± 0.0010
GaAs	Ga	2	80	295	0.0071 ± 0.0014	0.0040 ± 0.0005
GaAs	As	2	80	295	0.0056 ± 0.0011	0.0034 ± 0.0004
Ge	Ge	2	80	295	0.0058 ± 0.0008	0.0035 ± 0.0003
Ge (theory)	Ge	2	80	295	0.0055 ± 0.0002[a]	0.00338 ± 0.00004[a]

a. Values calculated from theory of Rehr et al.[13]

been made.[13] The experimental and theoretical results for the Debye-Waller factors of *Ge* are summarized in Table III verifying the validity of our fitting technique.

In comparing the experimental data to the theory, the relative energy origin E_o must be determined. Since the results were not sensitive to this energy origin we omit details of determining E_o here. Details can be found in Ref. 15.

Ratios of the experimental data to the theory are taken for the first shell, compensating for the Debye-Waller factors and the atomic absorption energy dependence as above. If the theory were correct, these ratios as a function of k would be constant at unity. As may be seen from Figs. 4a and 4b, there is a significant reduction of the first shell EXAFS relative to the theory. Furthermore, the reduction in amplitude is generally a function of k and varies with the system. The degree of amplitude reduction varies from about 40% for *CuBr* to 10-30% for *Ge*. As shown in Fig. 5, second shell

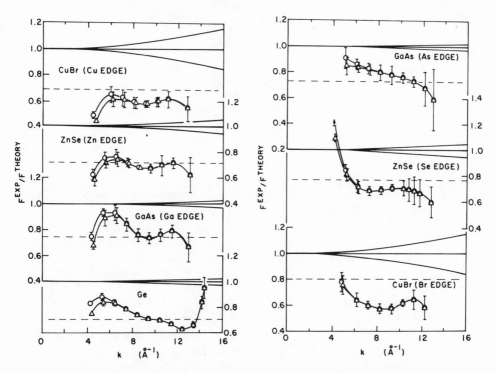

Fig. 4.   Ratios of first-shell data to theory for tetrahedral semiconductors after compensating for coordination number, $1/R^2$, and Debye-Waller factor. (O) inner potential chosen to match the experimental and theoretical phases at $k = 0$; (Δ) inner potential chosen to match the slope of the experimental and theoretical phases. The dashed lines are the atomic S_0^2 (from Reference 9) appropriate to each edge. The smooth curves above each figure correspond to a multiplicative error in both the data points and the error bars due to uncertainties in the Debye-Waller factor determination. The two extremes correspond to adding or subtracting the rms uncertainty from the most likely value. These errors were plotted separately because they reflect a shift of the entire ratio curve rather than random scatter.

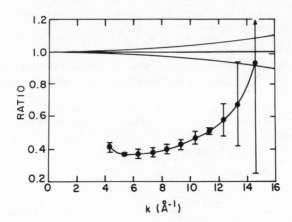

Fig. 5.   Ratio of second-shell to first-shell 80 K *Ge* data after compensating for coordination number, $1/R^2$ and Debye–Waller factor.

data for *Ge* exhibits a significantly larger reduction, in qualitative agreement with mean free path contributions to the amplitude.

More independent measurements may be made for the zincblende structure materials than for the diamond structure *Ge*. EXAFS was measured at the K edge of both constituents in each zincblende and the contributions of the first and second shells can be isolated about each type of atom. Because of the path lengths involved, the multiple-scattering peak with the lowest effective distance is about .5Å past the second shell peak, so contamination of the second shell data by multiple scattering is a negligible source of error. For a detailed discussion of these effects, see Lee and Pendry.[17]

We discuss the analysis of *CuBr* to illustrate the procedure followed for each zincblende sample. In this zincblende structure, if the center atom is a copper atom, the first shell contains 4 *Br* atoms at about 2.5Å and 12 *Cu* atoms at about 4.0Å, and vice versa for a *Br* center atom.

According to Eq. (1) the EXAFS amplitude depends only on the backscattering atom in the surrounding shells and not on the center atom. We can therefore determine the λ in two independent ways by taking the following ratios: $\dfrac{Cu(II)}{Br(I)}$ and $\dfrac{Br(II)}{Cu(I)}$, where $Cu(II)$ means the envelope of the EXAFS contribution of the second shell about *Cu* and similarly for the other terms. Thus the first ratio is that for backscattering *Cu* atoms in second to first shell and the second ratio is the same for *Br* atoms. According to Eq. (1) both of these ratios should equal $e^{-2(r_2 - r_1)/\lambda}$.

As can be observed from Fig. 6 where these two ratios are plotted, they are not equal. Clearly Eq. (1), the basic EXAFS expression, fails. We already have obtained evidence that Eq. (1) is in error by measurement[7] on Br_2 as discussed in Sect. II. These measurements showed the significant contribution of the many electron overlap factor S_0^2 implying that the $F(k)$ factor in Eq. (1) must be replaced by $S_0^2 F(k)$. This introduces the qualitatively new feature of the EXAFS amplitude depending not only on the type of backscatterer, but also on the type of center atom. Thus the two above ratios are not expected to be equal but, instead,

$$\frac{Cu(II)}{Br(I)} = S^2 e^{-2(r_2 - r_1)/\lambda} , \tag{3}$$

$$\frac{Br(II)}{Cu(I)} = \frac{e^{-2(r_2 - r_1)/\lambda}}{S^2} , \tag{4}$$

$$where \quad S^2 = S_0^2(Cu)/S_0^2(Br) .$$

We distinguish between the S_0^2 for Cu and for Br by adding the appropriate parentheses after the S_0^2. There are two other interesting ratios that can be formed from the measurements on the first two shells, namely,

$$\frac{Cu(II)}{Cu(I)} = T e^{-2(r_2 - r_1)/\lambda} \tag{5}$$

$$\frac{Br(II)}{Br(I)} = T^{-1} e^{-2(r_2 - r_1)/\lambda} \tag{6}$$

$$where \quad T = \frac{F(Cu)}{F(Br)} .$$

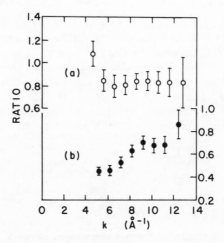

Fig. 6.   Ratios of second shell to first shell for 10 K $CuBr$, after compensating for coordination number, $1/R^2$, and Debye-Waller factor. (a) $Br(II)/Cu(I)$, (b) $Cu(II)/Br(I)$, where $A(n)$ is n^{th} shell data with the A atom excited. Note that in the simple theory, these two ratios should be the same.

The backscattering amplitude $F(k)$ for a Cu atom is denoted by $F(Cu)$ and similarly for a Br atom. From Eqs. (3) — (6) the 3 parameters λ, S^2, and T can be found as a function of k. The values of these parameters for $CuBr$ are plotted in Figs. 7(a), (b), and (c), respectively. Corresponding quantities are plotted in Figs. 8, and 9 for $ZnSe$ and $GaAs$, respectively, while λ is plotted for Ge in Fig. 10. Obviously, S^2 and T have no meaning for Ge.

V. Interpretation and Discussion

The k-dependence of λ for Ge illustrated in Fig. 10 follows the behavior expected for the electron mean free path as determined from auger electron data.[18] The mean free path measured by EXAFS differs somewhat from that of auger emission in that it is a convolution of the mean free path of the photoelectron and the lifetime of the core hole. However, since the core hole lifetime of Ge is much longer than the lifetimes corresponding to λ values of Ge plotted in Fig. 10, we note that λ is dominated by the photoelectron. Further evidence for the correctness of our assumptions is that the values of S^2 agree with theoretical estimates (dotted horizontal line in Figs. 7b, 8b, and 9b) within experimental error in the high k regime where the theoretical estimates are valid. The deviation at low k for $CuBr$ overlaps the regime where it is found from measurements[7] on Br_2 that the theory for S_0^2 does not apply.

The values for λ as determined from the tetrahedral series show a systematic variation increasing with increasing ionicity with its attendant increasing energy gap. This result is physically satisfying because a larger energy gap means a correspondingly larger energy to excite electron-hole pairs in the solid, which leads to less inelastic scattering of the photoelectron in the material.

At this point it is appropriate to point out that the mean free path concept has to be modified for the first shell. As discussed in Sec. II, the contribution to the inelastic scattering within the backscattering atom is already included in Lee's calculations of $F(k)$. In addition, inelastic energy losses in the center atom are accounted for by S_0^2.

In sum, then, λ effects are included in S_0^2 and the $F(k)$ calculations. For the first shell these should include a major portion of the λ effect. However, it is important to note that the mean free path between the first and second shells is not modified by S_0^2 and the $F(k)$ calculations. This is so because the *difference* in the energy loss mechanisms between the first and second shells is the loss suffered when the photoelectron travels the distance $(r_2 - r_1)$ and returns or $e^{-2(r_2 - r_1)/\lambda}$ as in Eqs. (3) — (6). The above argument and the results on the tetrahedral series indicate that the basic Eq. (1) based on the independent particle model must be modified as follows to include many-electron effects:

$$\chi(k) = \sum_i \frac{N_i S_0^2(k) F_i(k)}{k r_i^2} P_i(k) e^{-2(r_i - \Delta)/\lambda} \sin(2k r_i + \phi_i(k)) \qquad (7)$$

$S_0^2(k)$ accounts for the overlap effect in the center atom, and Δ accounts for the energy losses included by S_0^2 and the $F(k)$ calculation.

Figs. 7(c), 8(c), and 9(c) plot the various measured ratios of $F(k)$. These measured values are compared to the calculated values of Teo and Lee[3a] in these same figures (dotted curves). As can be seen, significant deviations occur between theory and experiment, especially in $CuBr$. Previous measurements[7] on Br_2 gave surprisingly

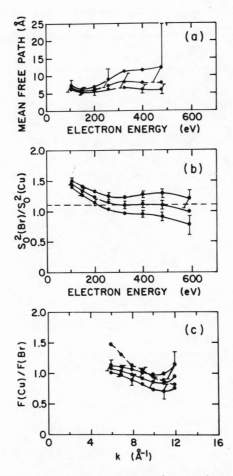

Fig. 7.   (a) Mean free path, (b) ratio of $S_0^2(Br)/S_0^2(Cu)$, and (c) ratio of $F(Cu)/F(Br)$ for $CuBr$. The three curves in each case correspond to error in the Debye-Waller factor determination where an rms error is added to or subtracted from the most likely value. The dashed line in the middle panel corresponds to the theoretical predictions of Ref. 9. The dashed line in the lower panel is the theoretical prediction of Ref. 3a.

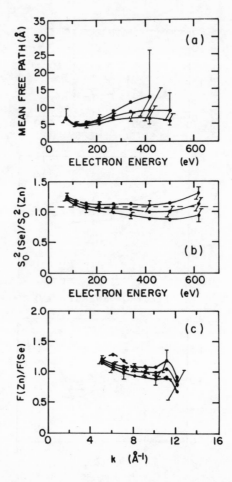

Fig. 8.   (a) Mean free path, (b) ratio of $S_0^2(Se)/S_0^2(Zn)$, and (c) ratio of
$F(Zn)/F(Se)$ for $ZnSe$. The three curves in each case correspond to errors
in the Debye-Waller factor determination where an rms error is added to or
subtracted from the most likely value. The dashed line in the middle panel
corresponds to the theoretical predictions of Ref. 9. The dashed line in the
lower panel is the theoretical prediction of Ref. 3a.

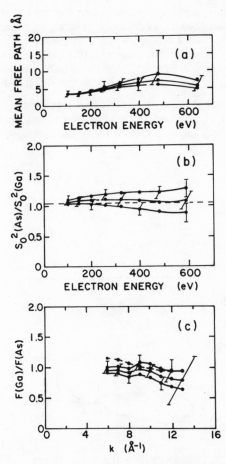

Fig. 9.   (a) Mean free path, (b) ratio of $S_0^2(As)/S_0^2(Ga)$, and (c) ratio of $F(Ga)/F(As)$ for $GaAs$. The three curves in each case correspond to errors in the Debye-Waller factor determination where an rms error is added to or subtracted from the most likely value. The dashed line in the middle panel corresponds to the theoretical predictions of Ref. 9. The dashed line in the lower panel is the theoretical prediction of Ref. 3a.

good agreement between Lee's calculations as corrected by S_0^2 and experiment. We therefore interpret the deviations in T for $CuBr$ as indicating that the Cu values for $F(k)$ are in error.

Values of Δ for the tetrahedral series can be estimated from the values of $[F(k)^{exp}/F(k)^{theo}]$ plotted in Fig. 4. The theoretical values of Teo and Lee were assumed correct for high k, since agreement is found at high k between the T ratios in Figs. 7(c), 8(c), 9(c) and theory. For Ge we have no experimental measurement to test the accuracy of Teo and Lee's calculation, but since all the neighboring atoms appear to be accurate, it is reasonable to assume that Ge is also accurate at high k. In all cases we find $\Delta \approx r_1$ as tabulated in Table IV. Finally, for 1st shell Cl backscattering in $CoCl_2$ and $MnCl_2$, we again find $\Delta \approx r_1$ which is also listed in Table IV.

The values of T shown in Figs. 7(c), 8(c), and 9(c) are the first direct comparison of the relative accuracy of Teo and Lee's calculation. The deviation found between theory and experiment may indicate some error in the approximations used in the calculations.

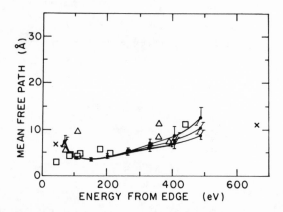

Fig. 10.   Mean free path for Ge. For comparison, results from Auger spectroscopy measurements are also given (from Reference 18). $X-Cu$, $\Delta-Ag$, $\square-A\ell$.

TABLE IV. Atomic overlap factors and core radii Δ for the dichlorides and the tetrahedral semiconductors. The values of S_0^2 are from reference 9. The nearest-neighbor distance is r_1.

	S_0^2	$F^{expt}/S_0^2 F^{theory}$	$[2(\Delta - r_1)/\lambda]$
$MnCl_2$	0.68	0.97 ± 0.12	-0.03 ± 0.12
$FeCl_2$	0.69	0.95 ± 0.14	-0.05 ± 0.15
$CoCl_2$	0.70	1.00 ± 0.11	0.00 ± 0.11
$CuBr$ (Cu edge)	0.70	0.86 ± 0.09	-0.15 ± 0.11
(Br edge)	0.80	0.83 ± 0.08	-0.19 ± 0.10
$ZnSe$ (Zn edge)	0.72	0.96 ± 0.07	-0.04 ± 0.07
(Se edge)	0.78	0.90 ± 0.12	-0.11 ± 0.13
$GaAs$ (Ga edge)	0.74	1.03 ± 0.07	0.03 ± 0.06
(As edge)	0.76	0.97 ± 0.05	-0.03 ± 0.05
Ge	0.75	0.93 ± 0.05	-0.07 ± 0.06

VI. Summary and Conclusions

The goal of this paper was to study transferability of amplitude which led us to consider many body effects in EXAFS. Two many body effects were discussed. The first is the decrease in the many electron dipole matrix element caused by the relaxation of the final state electrons not directly acted upon by the dipole operator. These passive electrons have an overlap S_0^2 with their corresponding initial states which is less than one because of their relaxation. As discussed, this leads to a decrease in the EXAFS at high energy of the order of 0.7 — 0.8 which value depends on the type of center atom. Thus the overlap introduces a new dependence of EXAFS on the center atom.

The second is the inelastic energy losses or lifetime effects which lead to the mean free path that is already included in the EXAFS expression, Eq. (1).

The study presented here showed directly that the basic EXAFS Eq. (1) has to be modified. We found that the many body modification of adding S_0^2 and $\Delta \approx r_1$ to produce Eq. (7) was adequate to fully understand our results. The many body effects produce corrections of the order of 20 — 30% to Eq. (1). However, if one employs standards which are similar to the unknown, most of the many body corrections cancel and it is possible to obtain much more accurate results just by using Eq. (1). In particular, the standard should have a center atom with the same S_0^2 value as the unknown (the same center atom automatically satisfies this requirement) and similar chemical environment and distances to the various shells. This is the same requirement we have been advocating for years.[4] The advance is that now we are obtaining insight into how to correct for variations between the standard and the unknown by use of Eq. (7). Since it is not possible to always find similar standards to the unknown, this new insight expands the possible applications of EXAFS while still maintaining the accuracy.

Our measurements indicate that there may be some error in the Lee calculations of $F(k)$ of Cu of the order of 20% at the lower k values. However, at the k values greater than 10Å^{-1}, the measurements and calculations agree. The discrepancy between the measurements and calculations of Cu could be due to the assumption of the small atom approximation[17,19] and/or the fact that in the tetrahedral sample the Cu and Br atoms transferred charge while the calculations assume a neutral atom. Calculations of $F(k)$ on atoms with charge transfer should be performed to check whether the discrepancy can be explained this way. Because of this k-dependent error in the calculations, errors will be realized in analysis methods which compare experiment directly to theory and interpret amplitude reduction in terms of a Debye-Waller factor.[20]

Finally, a more extensive study should be made to obtain an understanding of the factors that affect the magnitude of λ. For example, is the λ as measured by EXAFS the same as that of auger emission? One would expect that λ (auger emission) > λ (EXAFS) since the wave function coherence necessary for EXAFS is more severe than the auger emission requirement of a small energy loss. Can´a simple property of a material be correlated with λ, such as the energy gap?

Obtaining this further understanding will relax the requirement of using quite similar standards to the unknown in order to obtain high accuracy in EXAFS determination of structure.

Acknowledgments

The measurements reported here were made at the Stanford Synchrotron Radiation Laboratory (SSRL) which is supported by NSF grant #DMR77-27487 and the research is supported by NSF grant #DMR73-02521 A02. We would like to express our appreciation to Professor John Rehr for many helpful discussions.

References

1. E. A. Stern, Contemp. Phys. **19,** 289 (1978).

2. P. Eisenberger and B. M. Kincaid, Science **200,** 1441 (1978).

3. (a) B. K. Teo and P. A. Lee, J. Am. Chem. Soc. **101,** 2815 (1979); (b) P. A. Lee and G. Beni, Phys. Rev. **15,** 2862 (1977).

4. E. A. Stern, D. E. Sayers, and F. W. Lytle, Phys. Rev. **B11,** 4836 (1975).

5. E. A. Stern, Phys. Rev. **B10,** 3027 (1974).

6. C. A. Ashley and S. Doniach, Phys. Rev. **B11,** 1279 (1975).

7. E. A. Stern, S. M. Heald, and B. Bunker, Phys. Rev. Lett. **42,** 1372 (1979).

8. J. J. Rehr, E. A. Stern, R. L. Martin and E. R. Davidson, Phys. Rev. **B17,** 560 (1978).

9. T. A. Carlson, *Photoelectron and Auger Spectroscopy*, Plenum, N.Y. (1975).

10. T. A. Carlson and M. V. Krause, Phys. Rev. **140**, A1057 (1965); V. Schmidt, N. Sandner, H. Kuntzemüller, P. Dhez, F. Weilleumier, and E. Källne, Phys. Rev. **A13**, 1748 (1976); D. M. P. Holland, K. Codling, J. B. West and G. V. Marr, to be published.

11. T. A. Carlson, J. C. Carver, L. J. Saethre, F. G. Santibaney, and G. A. Vernon, J. of Electron Spect. and Related Phenom. **5**, 247 (1974).

12. W. H. McMaster, N. Kerr Del Grande, J. H. Mallett, and J. H. Hubell, *Compilation of X-ray Cross Sections*, National Technical Information Service, Springfield, Ba. (1969).

13. J. J. Rehr, H. Meuth, E. Sevillano, and S.-H. Chou (unpublished).

14. E. Sevillano, H. Meuth, and J. J. Rehr, to be published.

15. E. A. Stern, B. Bunker, and S. M. Heald, Phys. Rev. **B21**, 5521 (1980).

16. J. C. Phillips, Rev. Mod. Phys. **42**, 317 (1970).

17. P. A. Lee and J. B. Pendry, Phys. Rev. **B11**, 2795 (1975).

18. C. J. Powell, Surface Sci. **44**, 29 (1974).

19. R. F. Pettifer, 4th Eur. Phys. Soc. Gen. Conf., Chap. 7 (1979).

20. S. J. Gurman and J. B. Pendry, Solid State Commun. **20**, 287 (1976).

NEAR NEIGHBOR PEAK SHAPE CONSIDERATIONS IN EXAFS ANALYSIS

T. M. Hayes and J. B. Boyce

Xerox Palo Alto Research Center
Palo Alto, CA 94304

The extended X-ray absorption fine structure, or EXAFS, has become recognized as a reliable source of information about the atomic scale structure of complex molecules, liquids, and solids. Under the proper circumstances, one can determine not only the position but also the type and number of the nearest neighbors of the excited atom. Analysis of EXAFS data is complicated, however, by our inability to extract the low k components of the structural information. Determining the proper shape for the nearest neighbor peak in the pair correlation function becomes an essential, and frequently non-trivial, aspect of the analysis. Its neglect can lead to errors in determining the coordination number and, to a lesser extent, the mean nearest neighbor distance, as has been pointed out by Eisenberger and Brown.[1] On the other hand, using data on a superionic conductor, we demonstrate that EXAFS data of sufficient range and quality *will* enable a determination of the proper peak shape and, therefore, of the desired structural information. The incorporation of peak shape considerations into EXAFS analysis is discussed in general terms.

There are many nearly equivalent formulations for the EXAFS (see, for example, References 2-4). The following one[5-7] is the most convenient for this discussion because of the explicit appearance of the atomic pair correlation functions p. Consider the EXAFS on the absorption cross section for the photoexcitation of an electron from the K-shell of atom species α. Assume for simplicity that the nearest neighbors of α consist only of atom species β, distributed closely about a mean nearest neighbor distance r_1. The EXAFS due to these nearest neighbors alone can be represented approximately by

$$k \Delta_{\alpha\beta}(k) = (2\pi)^{\frac{1}{2}} r_1^{-2} \, 2Re \, [P_{\alpha\beta}(k) \, \Lambda_{\alpha\beta}(k,r_1)], \qquad (1)$$

where

$$P_{\alpha\beta}(k) = (2\pi)^{-\frac{1}{2}} \int\limits_{0}^{\infty} dr \, e^{2ikr} \, p_{\alpha\beta}(r) \qquad (2)$$

and

$$\Lambda_{\alpha\beta}(k,r) = -2i\,\pi^2\,t_\beta^+\,(-k,k)\,exp\,[-2\mu(k)r+2i\,\delta_\alpha(k)].\qquad(3)$$

In these expressions, Δ is the EXAFS (often labeled χ) expressed as a fraction of the K-shell absorption cross section and k is the final state electron momentum. Δ has been divided into two distinct contributions: structural information in P; complicated momentum dependences in Λ. $P_{\alpha\beta}(k)$ is the Fourier transform of the first peak in the radial distribution of atom species β about the excited species α, $p_{\alpha\beta}(r)$, defined so that $\int dr\ p_{\alpha\beta}(r)$ equals the number of nearest neighbor β atoms. In $\Lambda_{\alpha\beta}$ are included those factors which express the complex interactions of the final state electron with the excited atom (δ_α), the backscattering atoms (t_β^+), and the intervening media (μ).

Note from Equation 1 that the k range over which $k\Delta$ can be extracted from experiment is identical with the k range over which structural information (*i.e.*, P) can be obtained. $P(k)$ is related to the $S(q)$ of diffraction studies. The salient difference from the viewpoint of this discussion is that the variable complementary to r is $2k$ in the EXAFS case, not q as in the diffraction case. Accordingly, a comparison of the nature of the structural information to be obtained from the different techniques must compare the range of $2k$ in EXAFS to the range of q in diffraction.

The usual expression[2] for the EXAFS is obtained only for the idealized case where the peak in $p(r)$ is a Gaussian of half-width σ_1 representing a single shell of N_1 atoms at r_1:

$$k\Delta(k) = N_1\,r_1^{-2}\,|\Lambda(k,r_1)|\,exp\,[-2(\sigma_1 k)^2]\,2cos\,\{2kr_1+phase\,[\Lambda(k,r_1)]\}\,.\qquad(4)$$

It is unfortunate that Equation 4, or its equivalent, appears commonly in the literature, since it begs the question of peak shape. As will be shown, the peak shape may well not be Gaussian and this equation not appropriate.

The ultimate goal of most analyses of EXAFS data is to determine the nearest neighbor peak in the pair correlation function p (*i.e.*, the position, type, and number of the nearest neighbors). Unfortunately, Equation 1 is a single-scattering expression and cannot describe the oscillations in the absorption cross section for k below approximately 2.5Å^{-1} (a final state electron energy of 24 eV). There are two principal complications: (1) the cross section near the absorption edge is often dominated by localized resonance phenomena;[8] (2) the final state electrons have long mean free paths at low energies,[9] leading to the emergence of multiple scattering[4] as a significant factor in determining the EXAFS for k less than 2 or 3 Å^{-1}. As a result, the usual analysis, based on Equation 1, cannot be used to obtain structural information (*i.e.*, $P(k)$) from the low k region of an EXAFS measurement. The practical lower limit ranges from 2 to 3 Å^{-1} in typical situations. An effective upper limit occurs because of the decrease in both the scattering factor (t matrix) and $P(k)$ for high k. The value at which noise becomes a significant factor might vary from $k=9$ Å^{-1} for a light element backscatterer such as carbon to 20 Å^{-1} or more for a heavy element such as iodine.

As discussed above, the range of EXAFS data to be compared with a diffraction experiment is that of $2k$: approximately 5 to 30 Å^{-1}. In contrast, the information available from X-ray diffraction starts at $q=0$ and begins to deteriorate substantially in quality at 10 to 15 Å^{-1}. This important difference between EXAFS and diffraction data has been known for some time,[6] and was the subject of a short paper by Eisenberger and Brown.[1] The inability of EXAFS to yield structural information below $k=2.5$ Å^{-1} is a serious problem because it is $P(k)$ in precisely that k space region (*i.e.*, as k approaches zero) which is capable of yielding definitively the coordination number N and the mean nearest neighbor distance r_{nn}. Specifically, $P(0)$ is $(2\pi)^{-\frac{1}{2}}N$ and the wavelength of $P(k)$ as k approaches zero is π/r_{nn}. Since diffraction data approaches closely to $k=0$, N and r_{nn} may be determined in a straightforward manner. For example, a simple integral in r space leads to the proper coordination number. On the other hand, N and r_{nn} can be obtained from EXAFS data only through an extrapolation to $k=0$ from $k>2.5$ Å^{-1}. Any extrapolation requires some assumption about functional form — in this instance, the proper shape for the peak in the pair correlation function. The Gaussian shape which led to Equation 4 is commonly assumed in EXAFS analysis, but is often not the proper choice. In particular, some materials have highly asymmetric near neighbor peak shapes. In the following discussion, we demonstrate that it is possible to recognize those cases in which Gaussian is a poor approximation to the actual peak in $p(r)$ *if* the quality and the range of the data at high k are adequate. The proper functional form for the peak (*i.e.*, the appropriate structural model) can be determined by fitting the data, so that the extrapolation to $k=0$ to obtain the coordination number and the mean nearest neighbor distance will be meaningful.

We expect that many interesting candidates for EXAFS studies will exhibit nearest neighbor peak shapes which are non-Gaussian, including liquids, surfaces of solids, superionic conductors, anharmonic crystals, and, as a general class, materials at temperatures near or above their characteristic Debye temperatures (*e.g.*, zinc even at room temperature[1]). A superionic conductor, *CuI*, has been chosen to illustrate our approach because detailed analysis has demonstrated convincingly that only a strongly asymmetric peak shape can explain the EXAFS data on several of these materials (*i.e.*, *AgI*,[10,11] *CuI*,[12,13] *CuCl*,[12] and *CuBr*[12]) at a wide variety of temperatures. Furthermore, their unusual ionic conductivity has been shown to arise naturally from the same excluded volume model which yields this peak shape.[14] These EXAFS studies of superionic conductors are summarized in a later Chapter.

Consider the EXAFS due to the nearest neighbors of an excited Cu ion in $\gamma-CuI$ at 77 K and at 573 K, as shown in Figure 1.[12,13] These curves are obtained by Fourier transforming the data to r space, and then transforming the first peak *only* back into k space. The nearest neighbors are iodine ions in each case. The principal differences between these two spectra occur at high k, where the amplitude of the oscillations has decreased and the wavelength increased with temperature. This is consistent with the additional observation that the wavelength of the data at 573 K increases noticeably with k. We have been able to understand these features of the data in the context of an excluded volume model,[12,13] as follows. In the γ phase of CuI, the iodine ions are localized on a face-centered cubic (FCC) lattice. The $Cu-I$ pair correlation function is calculated in the excluded volume model by assuming a softened "hard sphere" repulsive potential for the $Cu-I$ interaction. This leads to a nearly uniform distribution of Cu ions in the crystal *except* that they are excluded from partially overlapping spheres, one centered at the locus of each iodine ion. The allowed region for Cu ions consists of tetrahedral and octahedral volumes, the latter having negligible occupancy at these temperatures.[13] Since the Cu ions must then occupy tetrahedral volumes, which lack a center of inversion symmetry, this model leads to a nearest neighbor peak in the radial distribution which is strongly asymmetric. Two symmetric models for the peak shape have also been considered: a single Gaussian where the amplitude has constrained to $N=4$ as expected for a tetrahedral site; a single Gaussian where the amplitude has been

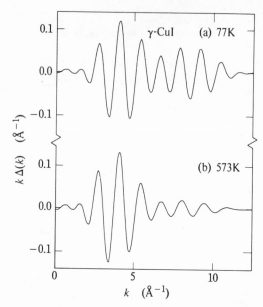

Fig. 1.   The EXAFS oscillations $k\Delta(k)$ from the nearest neighbors of Cu in $\gamma-CuI$ measured at 77 K (a) and at 573 K (b), after the contributions due to other than the nearest neighbor peak have been removed as described in the text.

adjusted to fit the data. Let us now compare the $k\Delta$ simulated using these models directly with the data in k space.

The sensitivity of the simulated $k\Delta$ to the shape of $p(r)$ is illustrated clearly in Figure 2. In Figure 2a is shown the nearest neighbor contribution to $k\Delta$ for CuI measured at 573 K (as shown in Figure 1b) together with the simulated $k\Delta$ corresponding to the best Gaussian fit with $N=4$ ($r_1=2.62$ Å, $\sigma_1=0.14$ Å). The residual error R calculated from a least squares comparison with the data in r space is 0.040.[12,13] The simulation is quite poor at high k. It is obvious that this Gaussian peak is unable to reproduce the k dependence of either the amplitude or the wavelength of the data.

In Figure 2b is shown the simulated $k\Delta$ for a Gaussian peak where N is allowed to vary from 4 to its favored value of 3.2 ($r_1=2.60$ Å, $\sigma_1=0.12$ Å). In this process, the value of R declines from 0.040 to 0.034, a significant improvement. By adopting a value of N which is physically unreasonable, this Gaussian model has achieved a better fit to the amplitude of the data at high k. If we had stopped at this point in the data analysis, we would have concluded erroneously that the Cu ions had found a new site where there are three iodine nearest neighbors. This is the pitfall discussed by Eisenberger and Brown.[1] Notice, however, that the fit to the phase of the oscillations in the data at high k has not improved. It is clear that further refinement in the model $p(r)$ is necessary. Part of the difficulty is that a $p(r)$ with a single Gaussian peak *cannot*

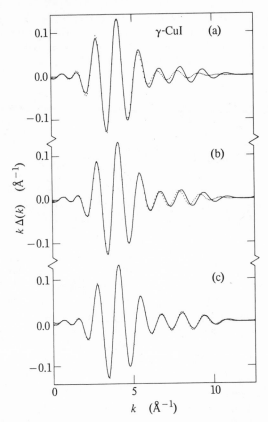

Fig. 2.   The EXAFS oscillations $k\Delta(k)$ from the nearest neighbors of Cu in $\gamma-CuI$
measured at 573 K (solid line), plotted together with three alternative simulations (dotted): (a) a single Gaussian with N fixed at 4; (b) a single Gaussian with N optimized at 3.2; (c) an excluded volume model with N fixed at
4. The last of these corresponds most closely with the data.

yield a variable wavelength in k space such as is found in the data. The fitting procedure will then lead to a simulation which fits the data only at low k, where the signal
is large. To the extent that the resulting simulation does not extrapolate properly to
$k=0$, the coordination number and the mean nearest neighbor distance obtained in the
fit will be misleading.

In Figure 2c is shown the simulated $k\Delta$ obtained from the excluded volume model,
where $N=4$. Since the R value has dropped substantially, to 0.009, a high confidence
level can be attached to the structural information deduced from this fit. Not only is
the amplitude of the data reproduced more closely for all k, but so is the phase. The
mean nearest neighbor distance r_{nn} is properly deduced to be the distance appropriate to

the tetrahedral Cu site, 2.64 Å. Inspection of the excluded volume $p(r)$ shown in Figure 3 reveals how this model is able to fit the variable wavelength observed in the data. Recall that the wavelength of $P(k)$ as k approaches zero is just π/r_{nn}. In contrast, the high k portion of $P(k)$, and therefore of $k\Delta$, is dominated by the sharpest feature in $p(r)$. In this case, that is the sharp rise at $r=2.43$ Å (which is less than r_{nn}). Since shorter r corresponds to longer wavelength in k space, the simulated $k\Delta$ will have a slightly longer wavelength at high k than at low k, just as is seen in the data. The phase information in the k space data has specified directly the principal features of $p(r)$: the wavelength at low k determined the mean nearest neighbor distance r_{nn}; the slightly longer wavelength at high k demanded a sharp rise in p at a value of r somewhat less than r_{nn}. In other words, $p(r)$ is unavoidably asymmetric in precisely the manner we have shown in Figure 3. Had we confined our analysis to Gaussian peak shapes, we would have underestimated not only the coordination number but also the mean nearest neighbor distance. In this instance, we were able to overcome the absence of structural information for $k<2.5$ Å^{-1} by demanding a truly good fit to the data over a wide range of k.

From this perspective, there is a clear disadvantage to a common practice of multiplying $k\Delta$ by k or k^2 prior to analyzing the EXAFS data. By weighting preferentially the high k portion of the spectrum, this procedure aggravates all of the problems discussed in this paper, making it still more difficult to obtain reliable values of N and r_{nn}.

Our study of superionic conductors was facilitated by prior knowledge of the number and symmetry of the nearest neighbors to be expected in the possible alternative sites for the cations. Even in cases where prior knowledge is minimal, however, the reasonableness of a Gaussian peak shape can be tested straightforwardly using simple func-

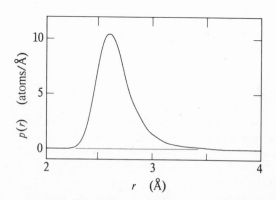

Fig. 3.   The pair correlation function $p(r)$ for the excluded volume model with N fixed at 4. The model parameters have been chosen to give an excellent fit to the nearest neighbor contribution to the EXAFS data from the Cu edge in $\gamma-CuI$ measured at 573 K.

tional alternatives. One possible set of functions might include $exp[-A(r-r_1)-B(r-r_1)^2]$, $exp[+A(r-r_1)^2-B(r-r_1)^4]$, etc. Eisenberger and Brown[1] have suggested a slightly different approach, which is convenient when the data is to be fit in k space: the $p(r)$ for an anharmonic potential is expanded in moments and Fourier transformed analytically to obtain a model expression for fitting in k space. The range of possible approaches is substantial, but all would have the same purpose: to test for *deviations* from a simple Gaussian shape. Once deviations are found, however, the analysis of those deviations will necessarily depend upon the nature of the system being studied. That is, a structural model consistent with the known properties of the system must be created and tested.

In summary, the absence of low k components from the structural information derived from EXAFS data is a significant handicap. In many interesting systems, EXAFS analysis must allow for the possibility that the shape of a peak in the pair correlation function might be other than a simple Gaussian. Failure to do so may lead to errors in determining the coordination number and, to a lesser extent, the near neighbor distances. On the other hand, we have shown that data of sufficient range and quality *will* enable a determination of the proper peak shape, leading to meaningful structural information. Finally, we have observed that there are several straightforward ways to test for deviations from Gaussian peak shapes. The challenge comes in devising the physical model which explains those deviations for the system in question.

ACKNOWLEDGMENTS

We gratefully acknowledge useful discussions with P. Eisenberger. The EXAFS data used in this work were measured at the Stanford Synchrotron Radiation Laboratory, which is supported by the National Science Foundation (under Contract DMR77-27489) in cooperation with SLAC and the Department of Energy.

REFERENCES

1. P. Eisenberger and G. S. Brown, Solid State Commun. **29**, 481 (1979).

2. E. A. Stern, Phys. Rev. B **10**, 3027 (1974).

3. C. A. Ashley and S. Doniach, Phys. Rev. B **11**, 1279 (1975).

4. P. A. Lee and J. B. Pendry, Phys. Rev. B **11**, 2795 (1975).

5. T. M. Hayes and P. N. Sen, Phys. Rev. Letters **34**, 956 (1975).

6. T. M. Hayes, P. N. Sen, and S. H. Hunter, J. Phys. C: Solid State Phys. **9**, 4357 (1976).

7. T. M. Hayes, J. Non-Cryst. Solids **31**, 57 (1978).

8. M. Brown, R. E. Peierls, and E. A. Stern, Phys. Rev. B **15**, 738 (1977).

9. See, for example, I. Lindau and W. E. Spicer, J. Elect. Spect. Rel. Phen. **3**, 409 (1974).

10. J. B. Boyce, T. M. Hayes, W. Stutius, and J. C. Mikkelsen, Jr., Phys. Rev. Lett. **38**, 1362 (1977).

11. T. M. Hayes, J. B. Boyce, and J. L. Beeby, J. Phys. C: Solid State Phys. **11**, 2931 (1978).

12. J. B. Boyce and T. M. Hayes in *Physics of Superionic Conductors*, ed. M. B. Salamon, Springer-Verlag, New York, 1979, Chapter 2.

13. J. B. Boyce, T. M. Hayes, W. Stutius, and J. C. Mikkelsen, Jr., Solid State Commun. **33**, 183 (1980).

14. T. M. Hayes and J. B. Boyce, Phys. Rev. B (in press).

DISORDER EFFECTS IN THE EXAFS OF METALS AND SEMICONDUCTORS IN THE SOLID AND LIQUID STATES

E. D. Crozier

Department of Physics
Simon Fraser University
Burnaby, B.C.
Canada V5A 1S6

INTRODUCTION

The EXAFS technique has been applied to structural determinations in a wide range of materials[1,2,3,4] with notable success. In the systems investigated the local structural order about the X-ray absorbing atom varied from the order of metallic single crystals[5] at 10K to the disorder[6,7] of amorphous semiconductors and metallic glasses. Usually, the structural analyses proceeded under the assumption, frequently implicit, that the distribution of nearest atoms was Gaussian, the effect of thermal and structural disorder being included through the Debye-Waller-like term $e^{-2k^2\sigma^2}$ in the EXAFS interference function $\chi(k)$. But for many systems a Gaussian distribution is only a first approximation.

In a crystal the assumption of a Gaussian distribution is equivalent to assuming that the lattice dynamics can be treated in the harmonic approximation. However, with increasing temperature, anharmonic contributions to the vibrational displacements of the atoms can no longer be neglected and effectively the distribution function becomes asymmetrical. The distribution function may also be asymmetrical in systems in which the packing of atoms is dominated by excluded volume effects, such as amorphous solids, superionic conductors[8] and liquid metals.[9] As indicated by Eisenberger and Brown,[10] the neglect of asymmetry effects will result in apparent decreases in the EXAFS deduced interatomic distances and coordination numbers and possible error in the identity of nearest neighbor species. The EXAFS analysis of systems with asymmetrical distributions becomes more complex. It requires the use of more elaborate models. Eisenberger and Brown have shown that the effects of small anharmonicity can be corrected. Methods of handling systems with more asymmetrical distributions have been discussed in the analysis of superionic conductors[8] and liquid, metals.[9]

In this paper results are presented for Ge, Se, As_2Se_3, Hg, Ga and Zn in the amorphous, crystalline or liquid states. The effects of increasing thermal and structural disorder are indicated. Methods of extracting structural parameters are discussed. Also discussed is the feasibility of using the high k information of EXAFS to parametrize the repulsive region of the ion-ion interaction potential of liquid metals.

Disordered Systems

The EXAFS interference function $\chi(k)$[11,12] can be written as

$$\chi(k) = \sum_j \frac{F_j(k)}{k} S_o^2(k) \int e^{-2r_j/\lambda} \sin(2kr_j + \delta_j(k)) \frac{P(r_j)}{r_j^2} dr_j \qquad (1)$$

where k is the wavevector of the photoelectron. The sum is over the atoms at the distances r_j from the X-ray absorbing atom, $F_j(k)$ is the magnitude of the amplitude for backscattering from the jth atom and $\delta_j(k)$ is an appropriate phase shift. The finite lifetime of the photoelectron final state is included through $e^{-2r_j/\lambda}$ and multielectron effects are included via $S_o^2(k)$.[13,14] The distribution of atoms is specified by $P(r_j)$, the probability that the jth backscattering atom is at r_j.

For the examination of highly disordered systems it is convenient to use a reduced EXAFS interference function $\chi'(k)$[9] defined in terms of the radial distribution function $g(R)$ of conventional diffraction studies. Assuming that the phase shift $\delta_j(k)$ and the factor $k^{-1}F_j(k)e^{-2r_j/\lambda}S_o^2(k)$ can be eliminated from $\chi(k)$, either by curve fitting a similar reference system or by using theoretical estimates, and recalling that $g(R)$ is defined such that $4\pi\rho g(R)R^2dR$ is the number of atoms in a spherical shell of radius R in a monatomic system of number density ρ, then (1) can be rewritten as

$$\chi'(k) = 4\pi\rho \int g(R)\sin 2kR dR . \qquad (2)$$

This equation provides an adequate approximation for the nearest neighbors. The functions $g(R)$ and $\chi'(k)$ are related by a sine transform as are $g(R)$ and the structure factor of diffraction studies

$$S(q) - 1 = 4\rho\pi \int (g(R) - 1) \frac{\sin qR}{qR} R^2dR \qquad (3)$$

where q, the momentum transfer, is twice the photoelectron momentum k. In principle, at least for the first shell of atoms, equivalent structural information should be contained in $S(q)$ or $\chi'(k)$ data which exist over equivalent regions of q or k space. However, the data do not span the same regions and consequently the distribution functions obtained by taking the inverse sine transform will differ.

As a specific example of the effects of the finite range of EXAFS data, consider the asymmetrical model function

$$g(R)_{asym} \begin{array}{ll} = A(R - R_o)^2 e^{-B(R-R_o)} & R \geq R_o \\ = 0 & R < R_o \end{array} \qquad (4)$$

and its corresponding reduced interference function

$$\chi'(k) = \frac{8\pi\rho A}{(B^2 + 4k^2)^3} \left[B(B^2 - 12k^2)\sin 2kR_o + 2k(3B^2 - 4k^2)\cos 2kR_o \right] \qquad (5)$$

where A, B and R_o are adjustable parameters.[9] If the inverse sine transform of Eq. 5 is taken then, provided the low k cut-off limit k_{min} equals zero and the upper cut-off k_{max}

is large, the original $g(R)_{asym}$ will be recovered as shown in Fig. 1a. Upon increasing k_{min}, the familiar truncation oscillations appear until as shown in Fig. 1b, in the EXAFS range, it is difficult to identify the correct peak in the sine transform.

This difficulty is avoided with symmetric distributions where the radial position can be determined from the position of the main peak in the magnitude of the complex transform[12]

$$\int_{k_{min}}^{k_{max}} k^n \chi(k) e^{-i(2kR+\delta(k))} dk.$$

However, for an asymmetric distribution, because of the inclusion of the cosine transform, the peak in the magnitude of the complex transform will not coincide with the position of R_1, the peak in $g(R)$, even for the unattainable case of $k_{min} = 0$ (Fig. 1a), and should not be used for the determination of radial distances. When the low k data is omitted the sine transform is also weighted towards small r values. For the asymmetrical model (Eq. 4) selected a typical EXAFS range of data produces a distance falling between R_1 and the cut-off distance R_o. If R_1 is desired, then it appears that it must be determined by curve fitting appropriate models.[8,9,10] On the other hand, the possibility exists[9] that the high k range available with EXAFS data can be used to advantage in highly disordered systems, such as liquid metals, in specifying the small R region of the effective two-body interaction potential.

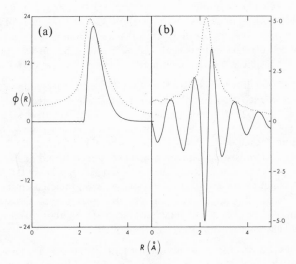

Fig. 1.   The Fourier transform of the reduced interference function $\chi'(k)$.[9] The sine transform — and the magnitude --- of

$$\int_{k_{min}}^{k_{max}} \chi'(k) e^{i2kR} dk$$

where $\chi'(k)$ is given by Eq. 5. In Fig. 1a, $k_{min} = 0.0 \text{Å}^{-1}$ and $k_{max} = 20 \text{Å}^{-1}$. In Fig. 1b, $k_{min} = 3.1 \text{Å}^{-1}$ and $k_{max} = 13.5 \text{Å}^{-1}$.

Since the presence of asymmetry in the distribution of atoms produces an error in EXAFS parameters it is essential to test all data sets for asymmetric effects. Eisenberger and Brown[10] have classified materials into classes which will or will not need asymmetry corrections depending on the magnitude of the disorder term σ_1^2. A sensitive test for the presence of asymmetry can be made by taking the complex transform

$$\phi(R) = \frac{1}{2\pi} \int_{k_{min}}^{k_{max}} k\chi(k)e^{i2kR} \left\{ \frac{e^{-i\delta(k)}}{Env(k)} \right\} dk \tag{6}$$

where $Env(k) = F_j(k)e^{-2r_j/\lambda} S_o^2(k)$. Both $\delta(k)$ and $Env(k)$ should be obtained by curve fitting procedures applied to a suitable reference material, although in the case of liquid Zn[9] it was adequate to neglect $S_o^2(k)$ and to approximate $F_1(k)e^{-2r_1/\lambda}$ by theoretical values.[16] If the peaks in the magnitude and imaginary parts of (6) coincide then the distribution function is symmetric, but if the peaks do not coincide then the distribution is asymmetric and corrections must be made.

If $A(k)$ and $A(k)_G$ represent the amplitudes of the interference functions of systems with an unknown and a Gaussian distribution, then the linearity of $\ell n(A(k)/A(k)_G)$ versus k^2 can be used to determine if the unknown distribution is Gaussian. However, in our experience the transform test is more definitive in simple systems with small asymmetric effects. It also appears to be less subjective than examining the k-dependence of the phase term for asymmetric contributions.

Semiconductors

Structural changes were examined in a temperature-dependent study of amorphous and crystalline Ge. Both amorphous and crystalline Ge have been studied before.[6,7,18] The present results involve higher temperatures and include data taken above and below the amorphous-crystalline transition temperature. The discussion will be more concerned with the possible effects of asymmetry than specific structural models.

A comparison of two different amorphous samples with a crystalline sample at 293K showed that the difference between the nearest neighbor distance for amorphous Ge and crystalline Ge, $R_1(a,293K) - R_1(c,293K)$, was 0.00 ± 0.01Å. This is in agreement with other EXAFS work,[17,18] but at possible variance with X-ray diffraction data where it was found[19] that $R_1(a,293K) - R_1(c,293K) = 0.02 \pm 0.02$Å. For amorphous Ge there was no consistent temperature dependence found in the EXAFS result for the nearest neighbor distance, $R_1(a,603K) - R_1(a,83K)$ being $0.0 \pm .01$Å. Over a similar range, thermal expansion data for crystalline Ge predicted an increase in R_1 of 0.009Å. On the basis of the EXAFS analysis it appears that the static structural disorder of amorphous Ge has not changed the nearest neighbor distance from that of crystalline Ge and that asymmetric effects are negligible.

At temperatures above the amorphous-crystalline transition asymmetric effects, while small, cannot be neglected. For example, at 1063K, the peaks in the magnitude and imaginary part of the transform (Eq. 6) did not coincide, the former indicating an apparent contraction of $0.04(4)$Å and the latter an apparent contraction of $0.02(6)$Å relative to the thermally expanded crystal at the same temperature. At $1063K$, which is 3.7 times the X-ray Debye temperature,[20] the motion of the Ge atoms is not expected to be harmonic. A simple one dimensional anharmonic oscillator model[9,10] was used to correct the data. The corrected distance was 0.002Å ± 0.01Å smaller than the expected thermally expanded value.

The temperature dependence of the disorder term of the nearest neighbors in amorphous and crystalline Ge is shown in Fig. 2. In the analysis it was assumed that the distributions of atoms was Gaussian causing $\chi_1(k)$ to vary as $e^{-2k^2\sigma_1^2}$. The difference, $\sigma_1^2(T) - \sigma_1^2(ref)$, between Ge at temperature T and a reference system of crystalline Ge at $293K$ was first obtained from the usual $\ln \chi_1(k,T)/\chi_1(k,ref)$ versus k^2 plots. An approximate value for the magnitude of $\sigma_1^2(ref)$ was then obtained by using theoretical values[16] to eliminate the factor $F_1(k)e^{-2r_1/\lambda}$ from $\chi_1(k)$ and curve-fitting the data only above k equal to 6.5Å^{-1} where the multi-electron term $S_o^2(k)$ was found to be approximately independent of k.[13,14] While the numerical procedures gave the differences $\sigma_1^2(T) - \sigma_1^2(ref)$ with a precision of $\pm 7\%$, the error in $\sigma_1^2(ref)$ is determined by the approximation made for $S_o^2(k)$ and may be greater than $\pm 20\%$. Additional error occurs in the case of crystalline Ge at the highest temperatures studied where the distribution of atoms is no longer Gaussian. However, it was estimated that the neglect of anharmonic contributions to the k-dependence of the amplitude of $\chi(k)$ led to effective values of σ_1^2 which were underestimated by less than 5%.

According to Eisenberger and Brown[10] the magnitude of the disorder term can be used as a guide in determining if asymmetric effects may be a problem. Specifically if $<x^2>/R_1$ and $<x^2>/\lambda < 0.01\text{Å}$ and if $(2k)^{2N+1} < x^{2N+1} > \ll 1$ then no correction is needed to maintain the 0.01 Å accuracy. Using the results of σ_1^2 of Fig. 2 as an approximation to $<x^2>$ it is found that the first two conditions are satisfied for amorphous Ge at $603K$ and crystalline Ge at $1085K$. It is difficult to estimate the magnitude of the third condition. In our analysis, asymmetry effects could be ignored in amorphous Ge but not in crystalline Ge at high temperatures.

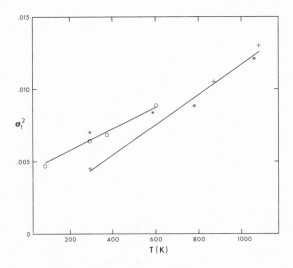

Fig. 2.   The temperature dependence of σ_1^2 of amorphous and crystalline Ge. The best straight lines are drawn for amorphous Ge, samples $A(*)$ and $B(\bigcirc)$ and crystalline Ge, samples $A(+)$ and $B(\diamond)$.

As expected σ_1^2 is greater for structurally disordered amorphous Ge than for crystalline Ge at the same temperature. The least-squares straight lines drawn in Fig. 2 indicate that σ_1^2 increases more slowly with temperature for amorphous Ge than for crystalline Ge. This could imply that a different force constant model is necessary for the two different states. However it is known that the physical properties of amorphous Ge change due to annealing as the crystallization temperature of $\sim$ 770 K is approached and that the changes can occur at temperatures as low as 450 K.[19] If the data for amorphous Ge, sample A, is examined only for temperatures less than 500 K it is found that the thermal dependence of σ_1^2 can be fitted to a simple Einstein model with an Einstein temperature $\theta_E = 360 \pm 11$ K. Crystalline Ge, sample A, can be fitted over the temperature range shown in Fig. 2 with $\theta_E = 351 \pm 10$ K. Thus it appears that the dynamic contribution to σ_1^2 in amorphous and crystalline Ge is determined by the same force constants.

Second and higher radial distances and coordination numbers are important in distinguishing between competing structural models for disordered systems. Normally with the EXAFS technique quantitative structural information is restricted to the nearest neighbors. However, it appears that σ_1^2 can be used to reject at least some models. For example, in an analysis of two fully relaxed continuous random network models involving clusters of 501 and 519 atoms, it was found that the static structural contributions to an assumed Gaussian disorder term were $.001\text{Å}^2$ and $.003$ Å^2 respectively.[22] In our work and that of Rabe et al[18] the increased disorder measured for amorphous Ge relative to crystalline Ge is sufficient to favor the 501 atom model over the 519 atom model. It is hoped that theoretical calculations of the thermal dependence of σ_1^2 in amorphous Ge which are being carried out by Rehr and co-workers[21] will permit a more sensitive test of specific models.

Se and As_2Se_3 were investigated in both the solid and liquid states. Below the melting point, σ_1^2 of Se was independent of temperature whereas above the melting point it increased linearly. In both solid and liquid As_2Se_3 σ_1^2 increased linearly with temperature although at a faster rate in the liquid state. Upon melting, a slight decrease occurred in the σ_1^2 obtained from an analysis of the K-edge EXAFS of both As and Se of As_2Se_3. In both systems at the highest temperatures studied the magnitudes of σ_1^2 were small. 78K and 146K above the melting points of Se and As_2Se_3 respectively σ_1^2 was only $\sim.010A^2$. Structural models for the systems will be discussed elsewhere.

In concluding this section, it is noted that the covalent bonding in semiconductors, such as Ge, Se and As_2Se_3, preserves a high degree of correlation in the motion of nearest neighbors. Asymmetry effects begin to contribute at the highest temperature studied, but such effects can be readily detected and may be possibly accounted for by relatively simple models.

Metals

The X-ray absorption spectra of the metals with the two lowest melting points, Hg and Ga, are shown in Fig. 3. In the solid state there is appreciable structure in the near edge and the extended regions of both the L_3 edge of Hg and the K edge of Ga. In liquid Ga the white line structure of the solid is retained and the EXAFS can be

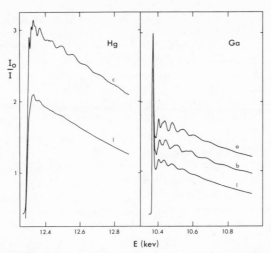

Fig. 3.   The X-ray absorption spectra of *Hg* and *Ga*. Polycrystalline *Hg* at 77K (c) and liquid *Hg* at 293K (l). A single crystal of *Ga* at 298K oriented with the polarization vector of the incident synchrotron radiation along the 2.47Å bond (a) and the 2.70Å bond (b). Supercooled liquid *Ga* at 298K (l). I_o/I is in arbitrary units and all curves have been displaced for clarity.

analyzed 100K above the melting point. In liquid *Hg*, even at the melting point, there is little structural detail in the near edge region and the amplitude of the EXAFS is almost zero by 12.4 kev. The rapid decrease in the amplitude of the *Hg* EXAFS is related in part to the energy dependence of the backscattering amplitude[16] which produces a pronounced minimum near 12.4 kev and, in part to the increased structural disorder of the liquid. In a comparison with diffraction studies it is observed that the oscillation in the structure factor $S(q)$ of liquid Hg[24] effectively vanishes by q equal to $12Å^{-1}$ which is close to 12.4 kev in the absorption spectrum. Whereas in liquid *Ga*, $S(q)$ still has structure[25] at $20Å^{-1}$ or 380 ev above the X-ray absorption *K*-edges.

In solids an estimate of the amplitude of the EXAFS expected at a given temperature can be obtained from knowledge of the Debye temperature. However, there is no simple relationship between the structural disorder of a liquid and a low Debye temperature. For example, the ratios of melting temperatures to Debye temperatures are 1.3, 2.3, 3.0 and 5.4 for *Ga*, *Hg*, *Zn* and *Se* respectively. Yet the EXAFS in the liquid state persists to high k in all but *Hg*.

The persistence of the EXAFS oscillation to high k values in liquid *Ga* may reflect a high degree of controversial local order. *Ga* is not a simple hard sphere fluid. Its structure factor has a shoulder on the high q side of the first peak, the origin of which is not fully understood. It has been frequently suggested that the anomalous structure of *Ga* reflects the presence of diatomic molecular-like associations, for example.[25] Recent theoretical calculations have attributed the structural anomalies to a dynamically screened fluctuating dipole interaction between ion cores.[26] Reference EXAFS parame-

ters, phase shift, effective backscattering amplitude and disorder term, which are necessary for a study of the liquid, cannot be extracted with accuracy from data on polycrystalline Ga because of beating between the first and second nearest neighbors at 2.47Å
and 2.70Å. Thus the EXAFS spectra of Fig. 3 on oriented single crystals were taken.
The analysis of the liquid is complicated by possible asymmetry effects as well as the
contributions of unusual local ordering. As indicated elsewhere in this book,[27] studies
in progress to examine the pressure-dependence of the small R region of the ion-ion
interaction potential in Ga should assist in clarifying the structure of the liquid.

   Zn has been the subject of extensive EXAFS investigations. In the case of oriented
single crystals of hcp Zn, Eisenberger and Brown[5,10] found a 5% contraction at 285K in
the nearest neighbor distance in the c-direction. They attributed that apparent contraction to anharmonic effects and the usual necessity in EXAFS analysis of discarding the
low k data. In a complementary study, polycrystalline Zn was examined at higher temperatures.[9] The nearest neighbor distance, in the a-direction, was found to display a
characteristic contraction. Asymmetry effects were readily identified by taking the
transform (6). As shown in Fig. 4, the distance determined from the peak position in
the sine transform displays a smaller contraction than that determined from the magnitude of the usual complex transform. A simple one dimensional analogue of the anharmonic model suggested by Eisenberger and Brown[10] provided reasonable agreement
with the thermal expansion data, giving at the highest temperature in the solid a
corrected R_1 value[9] which exceeded that expected from thermal expansion data by only
$\sim.02$Å. The model was used also to correct effective values of σ_1^2 which were obtained
from the data. At the highest temperature the corrected values were $\sim20\%$ higher than
those predicted by the Debye model of Beni and Platzman.

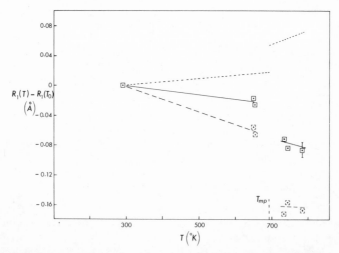

Fig. 4.   The temperature dependence of the nearest neighbor distance in Zn. $R_1(T)$
          estimated from thermal expansion data is indicated by ---, from the
          magnitude of the usual complex transform ● and from the sine transform
          defined in Eq. 6 ■. The solid — and dashed lines --- are drawn only as a
          visual aid.

Models for the EXAFS of Liquid Metals

The structure of simple liquid metals is determined approximately by the random packing of hard spheres. The nature of the effective ion-ion interaction potential cannot be totally neglected, particularly in EXAFS studies where the reduction in the amplitude of $\chi(k)$ at large k is sensitive to the softness of the repulsive potential. Model distribution functions for the nearest neighbors should be asymmetrical to account for the non-interpenetrability of the ion cores. The analytically simple function Eq. 5 was used as an empirical basis for curve fitting in k space a $\chi(k)$ which had been reduced without correcting for multi-electron effects. A 2 shell model gave the best fit.[9] The reconstructed $g(R)_{asym}$ is shown in Fig. 5 along with the $g(R)$ obtained from neutron diffraction data.[29] The need to scale $g(R)_{asym}$ is associated with an inadequate removal of the factor $F_1(k)e^{-2r_1/\lambda} S_o^2(k)$. However, the promising agreement suggests that simple functions such as Eq. 4 can be used to model the nearest neighbor distribution function of simple liquids.

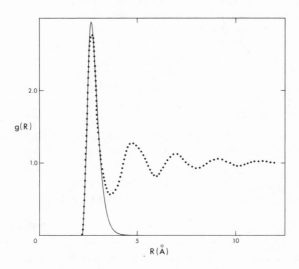

Fig. 5.   The radial distribution function for liquid Zn. The $g(R)$ obtained from neutron diffraction[28] at 733K is indicated by ... . The g_{asym} reconstructed from the k-space fit to the 2 shell asymmetric model (Eq. 5) is indicated by —. g_{asym} is scaled by the factor 0.75.

The Blip function theory of Jacobs and Anderson[30] was also applied to liquid Zn. In this theory the repulsive portion of the ion-ion interaction potential is treated as a perturbation of a hard-sphere fluid. Jacobs and Andersen have argued that the repulsive potential can be determined by forcing their Blip function $S(q)$ to fit the experimental $S(q)$ at large q. We used[9] the parameters specifying their model $S(q)$ to obtain a $g(R)$ which was then inverted via Eq. 2. The resulting Blip function reduced interference function is shown in Fig. 6. The same reduction procedure was also followed for an experimental X-ray diffraction $S(q)$.[31] The EXAFS $\chi'(k)$ was obtained by inverting the $\phi(R)$ obtained from Eq. 6, neglecting $S_o^2(k)$. Even with the inclusion of multielectron effects, it is evident that the amplitude of the Blip function $\chi'(k)$ is too large at large k relative to the EXAFS $\chi'(k)$, indicating that the model repulsive potential obtained by Jacobs and Andersen was not soft enough. This point is illustrated further by Fig. 7 which shows the potential of the mean force $U_m(R)$ obtained by inverting the defining equation[15] $g(R) = exp(-U_m/k_B T)$ where k_B is the Boltzmann constant.

CONCLUSIONS

The EXAFS analysis of thermally and structurally disordered systems can be complicated by the presence of asymmetry in the distribution of atoms. Since errors in structural parameters can result if asymmetric effects are not detected, all data sets should be tested for the presence of asymmetry. A transform analysis, at least for the simple systems discussed, provided a convenient and sensitive test.

In semiconductors, the covalent bonding preserves a high degree of local order. In Ge, Se and As_2Se_3 in the amorphous and crystalline states and in the liquid state, at least to the temperatures investigated, the effects of asymmetry were not appreciable and could be handled simply.

In pure metals, problems are encountered at lower temperatures. In solid Zn anharmonic effects in the nearest neighbor coordination in the a-direction were corrected at temperatures up to the melting point by using an anharmonic oscillator model. However, for the nearest neighbor in the c-direction, anharmonic effects could not be corrected above room temperature.[10] In liquid Hg the EXAFS was strongly attenuated by the structural disorder. In liquid Ga, which has a higher degree of structural order, analysis was complicated by the possible presence of asymmetric effects. In the case of liquid Zn, a simple empirical model was found to provide a promising reconstruction of the nearest neighbor distance.

Finally, it was shown that the high k values available with EXAFS data may be used to advantage in specifying the repulsive portion of the ion-ion interaction potential of liquid metals.

Thus, the loss of low k data may complicate the EXAFS analysis of high disordered systems, but important structural information can still be extracted.

ACKNOWLEDGEMENTS

I wish to thank A. J. Seary for helpful discussions and assistance with data analysis. I also wish to thank D. J. Kay, M. Plischtze and J. J. Rehr for useful discussions. This work was supported by grants received from the National Sciences and Engineering Research Council of Canada. The experimental EXAFS measurements were made at the Stanford Synchrotron Radiation Laboratory with the financial support of the National Science Foundation under contract DMR 77-27489 in co-operation with the Department of Energy, U.S.A.

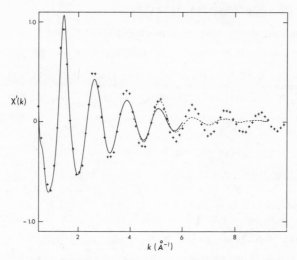

Fig. 6.   The reduced interference function $\chi'(k)$ of liquid Zn.[9] The EXAFS $\chi'(k)$ is given by ---, the Blip function by $+++$, and the X-ray diffraction by —.

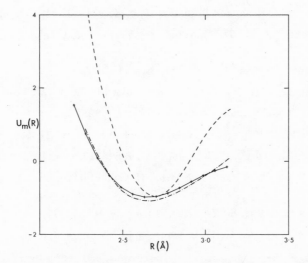

Fig. 7.   The potential of the mean force $U_m(R)$ in liquid Zn. $U_m(R)$ obtained by inverting the X-ray diffraction $g(R)$ is indicated by $+++$ and by inverting g_{asym} -·-. The potential used in constructing the Blip function $\chi'(k)$ of Fig. 6 is indicated by ---. Its ordinates have been displaced by an arbitrary amount.

REFERENCES

1. E. A. Stern, Contemp. Phys. **19**, 289 (1978).

2. P. Eisenberger and B. M. Kincaid, Science **200**, 1441 (1978).

3. D. R. Sandstrom and F. W. Lytle, Ann. Rev. Phys. Chem. **30**, 215 (1979).

4. S. P. Cramer and K. O. Hodgson, Prog. Inorg. Chem. **25**, 1 (1979).

5. G. S. Brown and P. Eisenberger, Solid State Commun. **24**, 201 (1977).

6. D. E. Sayers, in *Amorphous and Liquid Semiconductors (a-7)*, ed. W. E. Spear, G. G. Stevenson Ltd., Dundee, 1977, p. 61.

7. T. M. Hayes, J. Non-Cryst. Sol. **31**, 57 (1978).

8. (a) J. B. Boyce and T. M. Hayes, in *Physics of Superionic Conductors*, ed. M. B. Salamon, Vol. 15, Ch. 2, Springer, New York (1979); (b) J. B. Boyce, T. M. Hayes, W. Stutius and J. C. Mikkelsen, Jr., Phys. Rev. Lett. **38**, 1362 (1977); (c) T. M. Hayes, J. B. Boyce and J. L. Beeby, J. Phys. C. **11**, 2931 (1978).

9. E. D. Crozier and A. J. Seary, Can. J. Phys. **58**, 1388 (1980).

10. P. Eisenberger and G. S. Brown, Solid State Commun. **29**, 481 (1979).

11. E. A. Stern, Phys. Rev. **B10**, 3027 (1974).

12. E. A. Stern, D. E. Sayers and F. W. Lytle, Phys. Rev. **B11**, 4836 (1975).

13. J. J. Rehr, E. A. Stern, R. L. Martin and E. R. Davidson, Phys. Rev. **B17**, 560 (1978).

14. E. A. Stern, S. M. Heald and B. Bunker, Phys. Rev. Lett. **42**, 1372 (1979).

15. S. A. Rice and P. Gray, *The Statistical Mechanics of Simple Liquids*, Wiley and Sons, New York (1980).

16. B. K. Teo and P. A. Lee, J. Am. Chem. Soc. **101**, 2815 (1979).

17. D. E. Sayers, E. A. Stern and F. W. Lytle, Phys. Rev. Lett. **27**, 1204 (1971).

18. P. Rabe, G. Tolkiehn and A. Werner, J. Phys. C. **12**, L545 (1979).

19. R. J. Tempkin, W. Paul and G. A. N. Connell, Adv. Phys. **22**, 581 (1974).

20. B. W. Batterman and D. R. Chipman, Phys. Rev. **127**, 690 (1962).

21. J. J. Rehr, E. Sevillano, H. Meuth and S. H. Chou, Stanford Synchrotron Radiation Laboratory Report No. **79/05**, 32 (1979).

22. D. J. Kay, M.Sc. thesis, Simon Fraser University, B.C. (1978).

23. E. D. Crozier, F. W. Lytle, D. E. Sayers and E. A. Stern, Can. J. Chem. **55**, 1968 (1977).

24. P. J. Black and J. A. Cundall, Acta. Cryst. **19**, 807 (1965).

25. K. Suzuki, M. Misawa and Y. Fukushima, Trans. JIM. **16**, 299 (1975).

26. K. K. Mon, N. W. Ashcroft and G. V. Chester, Phys. Rev. **B19**, 5103 (1979).

27. R. Ingalls, J. M. Tranquada, J. E. Whitmore, E. D. Crozier and A. J. Seary, elsewhere in this book.

28. G. Beni and P. M. Platzman, Phys. Rev. **B14,** 1514 (1976).

29. W. Knoll, International Conference on Liquid Metals, 3rd, University of Bristol, 1976. Institute of Physics. Conference Series; No. 30, ed. R. Evans and D. A. Greenword, p. 117.

30. R. E. Jacobs and H. C. Andersen, Chem. Phys. **10,** 73 (1975).

31. Y. Waseda, *The Structure of Non-Crystalline Materials,* McGraw-Hill, New York (1980).

STRUCTURAL STUDIES OF SUPERIONIC CONDUCTION

J. B. Boyce and T. M. Hayes

Xerox Palo Alto Research Center
Palo Alto, CA 94304

EXAFS data on the normal and superionic phases of AgI and the cuprous halides have been analyzed using four structural models: harmonic oscillator, displaced site, anharmonic oscillator, and excluded volume. The most satisfactory description is obtained with the last model, based upon a softened hard-sphere pair potential. The results indicate that the tetrahedral locations in the halogen lattice are preferred by the mobile cations, but that at elevated temperatures substantial cation density also occurs at bridging trigonal sites, yielding the conduction path. Potential energy barrier heights are obtained. Finally, by modeling the conducting cations as a Boltzmann gas in the presence of the potential deduced from the EXAFS data, the temperature-dependent DC ionic conductivity is calculated.

Introduction

Superionic conductors are interesting in that they exhibit ionic conductivities comparable to those of molten salts while still in the solid phase (≈ 1 (Ω cm)$^{-1}$).[1] These materials include relatively simple binary salts like AgI[2-11] and structurally complex ceramics like $Li_2Ti_3O_7$.[12] Their ionic conductivities show different temperature dependences, ranging from predominantly exponential behavior with a sharp discontinuity at a phase transition (as in AgI) to a continuous exponential increase over a wide temperature range (as in $Na-\beta-Al_2O_3$ and $Li_2Ti_3O_7$).[13] The first category, dominated by Cu and Ag ion conductors, is particularly interesting since the onset of superionic conduction is associated with a true first order phase transition, accompanied by changes in structure and discontinuities in the specific heat. The entropy increase at these solid-solid transitions is often half the entropy increase on melting of a normal salt -- a fact which has led, in part, to the concept of sublattice melting at the transition.[14,15] AgI and the cuprous halides exemplify this behavior and are the materials we discuss here. Selected information on the phases is listed in Table 1.

103

Table 1. Selected information on the structures and phase
transitions of *AgI* and the cuprous halides.

Material	Phase	Transition Temperature (°C)	Structure
AgI	β		Wurtzite (Iodine HCP)
	α	147	Iodine BCC
	Melt	557	—
CuI	γ		Zincblende (Iodine FCC)
	β	369	Iodine HCP
	α	407	Iodine FCC
	Melt	600	—
CuBr	γ		Zincblende (*Br* FCC)
	β	385	*Br* HCP
	α	469	*Br* BCC
	Melt	488	—
CuCl	γ		Zincblende (*Cl* FCC)
	β	407	*Cl* HCP
	Melt	422	—

At all temperatures in these salts, the halogen anions are closely bound to their lattice sites and the cations are relatively mobile. As can be seen in Fig. 1, the ionic conductivity at low temperatures is like that of a normal salt. That is, it is low $(<10^{-4} \, (\Omega \, cm)^{-1})$ and is highly activated $(U \approx 1 \, eV)$. In this normal phase, the cations are situated at well defined lattice sites and have very low mobility. They become much more mobile as the phase transition is approached. In *AgI* the conductivity increases abruptly by a factor of 10^4 at the $\beta \rightarrow \alpha$ phase transition.[16,17] The situation for the cuprous halides[18] is different in that the conductivity increases more rapidly than exponential and achieves a large value $(\approx 0.1 \, (\Omega \, cm)^{-1})$ *below* the first phase transition. It increases further by a small factor *at* the two phase transitions, $\gamma \rightarrow \beta$ and $\beta \rightarrow \alpha$. In each case, however, the α phases exhibit superionic behavior. That is, the cationic conductivity is large $(\approx 1 \, (\Omega \, cm)^{-1})$ and only slightly temperature dependent $U \approx 0.1 \, eV)$. (Although *CuCl* has no α phase, its conductivity is as substantial as that of *CuBr* and *CuI* at the high temperature end of the γ phase and in the β phase.[18])

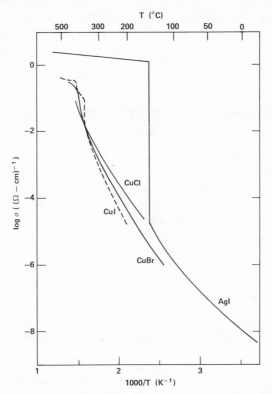

Fig. 1   Conductivity on a log scale versus inverse temperature for AgI [16,17] and the
cuprous halides. [18] Cationic conductivity dominates over this temperature
region.

    The characterization of these materials requires understanding the nature of the
transitions to the superionic phase and of the motion of the cations that accounts for
the high conductivity in this phase. Structural studies are particularly interesting in this
regard. By yielding the location of the ions, they can reveal both the conduction path
and the nature of the cation disorder in the superionic phase. The structural information probed through the extended x-ray absorption fine structure (EXAFS) is especially
useful in this respect. Since the EXAFS on the cation K-shell absorption measures primarily the pair correlation function of the mobile cations with respect to the immobile
anions, this technique is especially well suited to determining the path taken by the
conducting ions. In the following, we shall summarize the results of EXAFS studies of
AgI, CuI, $CuBr$, and $CuCl$. It will be seen that the atomic scale information deduced
from the EXAFS can yield valuable insight into the cation distribution and the conduction mechanism. Specifically, the data are analyzed in terms of a three-dimensional
one-ion potential which determines the cation positions. In addition, by treating the
conducting ions as a Boltzmann gas in the presence of this potential, the probability
that an ion will progress from one lowest energy site to another can be calculated. This

leads to an estimate of the temperature-dependent DC ionic conductivity which correlates strongly with experimental determinations in AgI[16,17] and in the Cu halides.[18] Analysis of the EXAFS data has led directly to a model which can not only explain the structural data but can also yield substantial insight into the details of the unusual ionic conduction in these materials.

Data Analysis Methods

The EXAFS data on the superionic conductors[19] will be discussed using the real space representation. The Fourier transform of the EXAFS on the K-shell absorption of atom species α may be written as[20]

$$\phi_\alpha(\mathbf{r}) = \sum_\beta \int_0^\infty d\mathbf{r}'/\mathbf{r}'^2 \, p_{\alpha\beta}(\mathbf{r}') \, \xi_{\alpha\beta}(\mathbf{r}-\mathbf{r}') \ . \tag{1}$$

The EXAFS is divided into two parts. The structural information is in $p_{\alpha\beta}(\mathbf{r})$, while the details of the electron scattering processes are contained in the peak function $\xi_{\alpha\beta}(\mathbf{r})$. $p_{\alpha\beta}(\mathbf{r})$ is the pair correlation function of β atoms about the excited atom α. It is normalized so that $\int d\mathbf{r} \, p_{\alpha\beta}(\mathbf{r})$ is the total number of β atoms in the sample. The sum over β in Eq. 1 includes all the atom species in the sample. In essence, ϕ is a linear combination of ξ's, one located at the position of each peak in p. Figure 2a shows an example of $\phi(\mathbf{r})$, derived from the Cu K-shell absorption in CuI at 77K. The double-peaked structure at approximately 2.2 Å is the signal from a single shell containing the four iodine neighbors to each Cu atom in this zincblende structure. ξ_{Cu-I} is double-peaked due to a Ramsauer-Townsend[21] resonance in the iodine backscattering t-matrix. The next two peaks are due to the Cu second neighbors and the iodine third neighbors, respectively. The first neighbor peak dominates the spectrum principally because further neighbors correspond to much broader peaks in $p(\mathbf{r})$.[22] The positions of the peaks in Fig. 2a are shifted inward from the actual near neighbor spacing because $\xi(\mathbf{r})$ peaks at $\mathbf{r} < 0$.

The first peak in $\phi(\mathbf{r})$ is resolved from the others and so can be analyzed separately. It is the one of interest for the studies presented here since it yields the location of the mobile cations relative to the immobile anion lattice. In fact, only the first peak is observed with adequate signal-to-noise at elevated temperatures, as can be seen in Fig. 2b and c. These are the real space data on the Cu K-edge of CuI at 300°C in the high temperature end of the γ phase (Fig. 2b), and at 470°C in the superionic α phase (Fig. 2c). The sample and the data reduction procedure are the same as for Fig. 2a. The $\phi(\mathbf{r})'s$ differ substantially, however, with the first neighbor peak being broader and reduced in amplitude at the higher temperature. In addition, it has shifted inward by about 0.12 Å despite the fact that lattice expansion has caused an increase of 0.03 Å in the $Cu-I$ near neighbor spacing. Since ξ is insensitive to crystal structure, local bonding, thermal effects, etc.,[23,24] the changes that are observed between Figs. 2a, b, and c are due entirely to changes in $p(\mathbf{r})$ with temperature.

We quantify these changes using Eq. 1 in the following manner. Since $p_{\alpha\beta}(\mathbf{r})$ is a narrow Gaussian of known position at low temperatures in these materials, the unchanging ξ may be extracted using Eq. 1 from the $\phi(\mathbf{r})$ measured at 77K. This $\xi(\mathbf{r})$ can then be used to determine the unknown $p(\mathbf{r})$ at elevated temperatures. This is accomplished by formulating a model for $p(\mathbf{r})$. From the model $p(\mathbf{r})$ and the $\xi(\mathbf{r})$ obtained from the low temperature data, one can calculate a model $\phi_m(\mathbf{r})$ using Eq. 1. One can then compare $\phi_m(\mathbf{r})$ with the data and adjust the model parameters using a

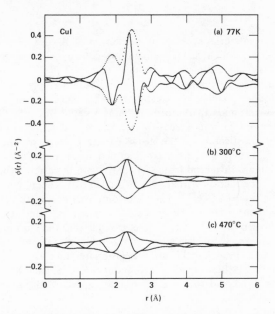

Fig. 2   The real part (solid line) and the magnitude (dotted line) of the Fourier transform of the EXAFS on the *Cu K*-edge in *CuI* (a) at 77K, (b) at 300°C inthe high temperature end of the γ phase, and (c) at 470°C in the superionic α phase. The window used for the transform is $k = 2.85$ to 13.6 $\overset{\circ}{A}^{-1}$, Gaussian broadened by 0.7 $\overset{\circ}{A}^{-1}$.

least squares criterion. The parameter which is minimized is

$$R = (2N)^{-1} \sum^{N} \{Re\,(\phi-\phi_m)^2/[(Re\ \phi)^2 + (Re\ d\phi/d\mathbf{r})^2]$$

$$+ Im\,(\phi-\phi_m)^2/[(Im\ \phi)^2 + (Im\ d\phi/d\mathbf{r})^2]\} \ , \qquad (2)$$

where the sum is over the N data points in the range of the spectral feature of interest. This is a weighted fractional difference least squares, where the $d\phi/d\mathbf{r}$ term has been added to enhance the sensitivity of R to variations in the shape of $p(\mathbf{r})$ at the expense of some of its great sensitivity to position. This procedure yields an optimal set of parameters for a particular model $p(\mathbf{r})$. One then repeats this process for other physically realistic models for $p(\mathbf{r})$. If the various models yield R values which are substantially different, then one may conclude that the EXAFS contains sufficient information to choose among the structural models. This is the case for the superionic conductors as will be discussed below.

Structural Models

To provide a frame of reference for the discussion of the EXAFS data, we first discuss the various models which have been proposed to explain structural data on the superionic conductors. The simplest approach is to model $p(\mathbf{r})$ directly. For example, it could be assumed that the nearest neighbor contribution to $p(\mathbf{r})$ is a single Gaussian peak: a harmonic model. Such a $p(\mathbf{r})$ results from the harmonic approximation to the ion potential, and corresponds to the cations being located in a site that is equidistant from the neighboring anions and executing symmetric, harmonic vibrations about this site. Slightly more sophisticated are the displaced site models, in which the cations in the high temperature superionic phase are assumed to be statistically distributed over metastable sites which are displaced from the central site in certain symmetry directions (*e.g.*, along probable conduction paths). Such models have been applied to AgI,[5,6,8,11] CuI,[25–27] $CuBr$,[26,28,29] and $CuCl$.[29–31] They are somewhat successful since they yield an asymmetric near neighbor pair correlation function, which is needed to fit the structural data. They can be criticized, however, on the grounds that they are phenomenological, static, and often yield small displacements comparable to vibration amplitudes.[6] One would expect a dynamic model to be more appropriate since the ions are mobile. In addition, we have tested these models using our EXAFS data on AgI[19,32] and the cuprous halides,[19,27,29] and find they are inferior to an excluded volume model, as will be discussed.

A more general approach to formulating a structural model begins by defining a microscopic potential $V(\mathbf{r})$ in which the cations move. The ion density follows directly from this potential through the Boltzman relation:

$$\rho(\mathbf{r}) = \rho_0 \, exp\,(-V(\mathbf{r})/k_B T) \ . \tag{3}$$

The problem of modeling the structure has become one of modeling the potential. It is possible to use a very general form for the ion-ion pair interaction, which would include the core-core repulsion, Coulomb interaction, and ionic polarizabilities. This approach has been taken in various calculations, including the molecular dynamics simulations[33] and calculations of the potential energy of ions of varying size in the $\alpha-AgI$ lattice by Flygare and Huggins.[34] It is very cumbersome, however, in analyzing structural data, leading to the use of much simpler models for the potential. We will now discuss two of these, the anharmonic and excluded volume models.

In the anharmonic models, the ions are assumed to be independent oscillators and the potential is expanded about an equilibrium position. This approach yields ionic potentials of the form

$$2V(\mathbf{r}) = V_0 + \tfrac{1}{2}\,a\,\mathbf{u}^2 + b\ xyz + \cdots , \tag{4}$$

where $\mathbf{u} = (x,y,z)$ is the displacement of the ions from this equilibrium site and V_0, a, and b are constants. The precise form of the cubic and higher order terms in Eq. 4 depends on the lattice symmetry. The form shown is for tetrahedral symmetry, that is appropriate, for example, for Cu in $\gamma-$ and $\alpha-CuI$. If the anharmonic terms are negligible, then Eqs. 3-4 lead to a Gaussian pair correlation function with a width $\sigma \approx (k_B T/a)^{1/2}$. If the anharmonic terms are non-negligible, the pair correlation function becomes asymmetric, similar to the shape required by the structural data. Therefore, this model has been somewhat successful for AgI,[6,9,11] CuI,[27,35] $CuBr$,[29,36] and $CuCl$.[29-31,37] One expects an expansion about the equilibrium site of the form of Eq. 4 to break down, however, for the superionic conductors where a cation must leave one equilibrium site and hop to another. An indication that this is the case is found in the neutron diffraction work on AgI by Cava et al,[9] who find that significant fourth order terms must be included in the anharmonic expansion of the potential. In addition, we have tested the anharmonic model, with cubic terms only, using EXAFS data on the cuprous halides,[19,27,29] and find it to be inferior to another approximation to the potential, the hard sphere approximation which we now discuss.

The excluded volume model for $V(\mathbf{r})$[32] assumes that the hard core repulsion dominates the ion-ion interaction. The anions are fixed at their lattice sites and only cation-anion interactions are considered. The cation density is then a constant ρ_0 *except* that it is zero within a distance of $\mathbf{r}_{excluded}$ of an anion site, where $\mathbf{r}_{excluded} = \mathbf{r}_c + \mathbf{r}_a$, the effective hard sphere radii of cation and anion. For one anion at the origin, the cation density $\rho_c^1(\mathbf{r})$ is given by

$$\rho_c^1(\mathbf{r}) = \begin{cases} 0, \ \mathbf{r} < \mathbf{r}_{excluded} \\ \\ \rho_0, \ \mathbf{r} > \mathbf{r}_{excluded} \end{cases} . \tag{5}$$

While Eq. 5 is appropriate for a hard core interaction, the actual interactions have a finite slope, or softness, and the anions are known to execute large amplitude vibrations about their lattice sites. To account for these effects we convolute Eq. 5 with a Gaussian of width σ. The density can then be expressed in terms of the complementary error function as

$$\rho_c^1(\mathbf{r})/\rho_0 = \begin{cases} erfc(x)/2, \ \mathbf{r} < \mathbf{r}_{excluded} \\ \\ 1 - erfc(x)/2, \ \mathbf{r} > \mathbf{r}_{excluded} \end{cases} , \tag{6}$$

where $x^2 = (\mathbf{r} - \mathbf{r}_{excluded})^2/2\sigma^2$. To obtain the total cation density, $\rho_c(\mathbf{r})$, one must include the effect of all the anion near neighbors, located at $\mathbf{r}_j$. Eq. 6 is then generalized to yield

$$\rho_c(\mathbf{r}) = \rho_0 \, \Pi_j^N \, \rho_c^1(\mathbf{r} - \mathbf{r}_j)/\rho_0 \ , \tag{7}$$

where the product includes all the N anion sites in the crystal. From this cation density, the cation-anion pair correlation function $p_{c-a}(\mathbf{r})$ can be calculated:

$$p_{c-a}(\mathbf{r}) = 4\pi \mathbf{r}^2 N^{-1} \sum_j^N <\rho(\mathbf{r} + \mathbf{r}_j)>_{\Omega_r} \ . \tag{8}$$

In fitting the EXAFS data using Eq. 8, the lattice of the anions is determined from diffraction results. Only the two parameters $\mathbf{r}_{excluded}$ and σ need be adjusted to determine the cation-anion pair correlation function in many systems.

In the FCC and HCP phases of the cuprous halides, however, an additional parameter must be added. One can distinguish two types of voids for the *Cu* ions, tetrahedrally and octahedrally coordinated sites. In the FCC halide lattice at low temperatures, only alternate tetrahedral voids are occupied. As the temperature is increased, however, the octahedral occupation is expected to increase since the conduction path for motion through the faces of the polyhedra will consist of alternating tetrahedral and octahedral sites.[38] Since the near neighbor environment is different for *Cu* ions in these two sites, their relative occupation will be determined by interactions in addition to the softened hard sphere repulsion. These can be treated phenomenologically by superimposing on $V(r)$ a square well potential which is zero in the tetrahedral location and greater than zero in the octahedral location. Accordingly, a third parameter is introduced in fitting the cuprous halide data, the fraction of *Cu* ions found in the octahedral sites c_{oct}. The near neighbor $p_{c-a}(r)$ for these materials will consist of the sum of two parts: a tetrahedral contribution with four halide ions around a *Cu* ion, weighted by $c_{tet} = (1 - c_{oct})$; an octahedral contribution with six halide ions around a *Cu* ion, weighted by c_{oct}. The three parameters, $r_{excluded}$, σ, and c_{oct}, are adjusted to fit the data. We now consider the experimental results on these superionic conductors, and the ability of the various models to explain the data.

Discussion of AgI Data

The EXAFS on the *Ag* K-edge in *AgI* was measured as a function of temperature from 77K in the low temperature β phase up into the superionic α phase.[8,32] The 77K data on β-*AgI* served as the structural standard. At this temperature the *Ag* ions are at the center of an iodine tetrahedron in the wurtzite structure, so that the near neighbor *Ag*$-I$ pair correlation function, $p_{Ag-I}(r)$, is a narrow Gaussian centered at 2.81 Å with $N=4$ neighbors. The width is somewhat uncertain but is typically ≈ 0.06 Å. Various structural models for the superionic α phase were tested. We summarize the results using the data on α-*AgI* at 198°C:

(1) For a single Gaussian harmonic model, the number of iodine neighbors N, the *Ag*$-I$ nearest neighbor spacing, r_0, and the Gaussian width, σ, were adjusted to fit the data by minimizing R. This model did not fit very well, yielding $R=4.5\%$ with $r_0 = 2.81$ Å and $N=2.72$ neighbors (see Table 2). Although the value of r_0 is very close to the value of 2.83 Å for the center of the tetrahedral site, the amplitude is reduced from the 4 expected for that site.

(2) The Strock model[2] distributes the two *Ag* ions uniformly over the 42 crystallographic sites in the unit cell formed by the BCC iodine ions. These 42 sites consist of 12 tetrahedral, 6 octahedral and 24 trigonal locations. $p_{Ag-I}(r)$ in this case is a sum of Gaussians corresponding to each of the three types of sites, weighted by the degeneracy of the sites. The Gaussian width was adjusted for a best fit to the data. This model did not fit well, yielding $R=11.9\%$.

(3) For the displaced site model, displacements in two different directions from the tetrahedral site have been considered: $<100>$ and $<110>$.

Table 2. The parameters obtained for several structural models when adjusted to fit the data on superionic AgI and CuI. The uncertainty in the determined distances is 0.01 Å and in the other parameters is $\approx 10\%$.

Material	Model	R (%)	N	r_0 (Å)	σ (Å)
$\alpha-AgI$	Tet site		4	2.83	
198°C	Harmonic	4.5	2.7	2.81	0.13
	Strock	11.9	(see text)		
	Displaced site:	4.2	1.28	2.74	0.09
	$<100>$		1.28	2.88	0.09
	Displaced site:	3.1	1.95	2.77	0.10
	$<110>$		0.65	2.93	0.10
	Excluded volume	2.0	$[r_{excluded}=2.69]$		
$\alpha-CuI$	Tet site		4	2.67	
470°C	Oct site		6	3.08	
	Harmonic	6.9	4	2.59	0.20
		3.5	2.2	2.57	0.14
	Displaced site:	7.9	3	2.54	0.17
	$<111>$		1	3.21	0.27
		4.2	3	2.59	0.17
			1	3.15	0.29
	Anharmonic	4.6	(see text)		
	Excluded volume	1.8	$[r_{excluded}=2.44; c_{oct}=30\%]$		

(a) For a $<100>$ displacement, the Ag ions are moved from the tetrahedral center toward octahedral sites. $p_{Ag-I}(\mathbf{r})$ is the sum of two Gaussians, centered at $\mathbf{r}_1$ and $\mathbf{r}_2$ and each having the same amplitude ($N=1$) and the same width. Buhrer and Halg[5] proposed such a model with a displacement of 0.29 Å at 195°C corresponding to $\mathbf{r}_1 = 2.72$ Å and $\mathbf{r}_2 = 2.97$ Å.

This particular displacement did not fit the 198°C EXAFS data ($R=13.1\%$). By allowing $\mathbf{r}_1$ and $\mathbf{r}_2$ to vary, the fit improved somewhat but still did not reproduce the data well. If the amplitude is allowed to vary in addition, the fit improves substantially to $R=4.2\%$ for the parameters shown in Table 2. The reduced amplitude (2.56 neighbors in all instead of 4) and the rather high R indicate that this model does not yield a good approximation to $p_{Ag-I}(\mathbf{r})$.

(b) For a $<110>$ displacement, the Ag ions are moved from the center of the tetrahedral sites toward the faces of the tetrahedra, the trigonal sites. This is a likely displacement for Ag ion conduction between tetrahedral locations, through the shared faces.[38] In this case, $p_{Ag-I}(\mathbf{r})$ is the sum of two Gaussians with amplitudes in the ratio of 3 to 1[8] (since this displacement moves an Ag ion closer to the three iodine ions of the tetrahedral face and further from the one iodine ion at the tetrahedral corner). If the amplitudes were fixed at $N_1 = 3$ and $N_2 = 1$, then the fit was not good. If the amplitudes were allowed to vary with the ratio fixed at 3:1, then the fit improved substantially to $R=3.1\%$ with the parameters shown in Table 2. This fit corresponds to a reduced amplitude of 2.60 neighbors instead of 4.

(4) In the excluded volume model, the Ag ions are distributed uniformly over all the allowed space in the iodine BCC lattice, being excluded only from a region about the iodine ions described by the softened hard-sphere interaction.[32] The largest such allowed void is the tetrahedral location. The amplitude of $p_{Ag-I}(\mathbf{r})$ is completely determined in this model since one includes all four iodine nearest neighbors in Eq. 8. Specifying that the iodine ions form a BCC lattice with $a_0 = 5.068$ Å at $200°C$,[5] the two parameters $\mathbf{r}_{excluded}$ and σ were adjusted for a best fit to the EXAFS data at 198°C. This model yielded a good fit with $\mathbf{r}_{excluded} = 2.69$ Å and $\sigma = 0.08$ Å $(R=2.0\%)$, substantially better than the results of the other models considered $(R>3.1\%)$. Similarly good fits were obtained for the data at all the other temperatures studied and resulting values of $\mathbf{r}_{excluded}$ are given in Fig. 3a. Also shown is $\mathbf{r}_{face}$, that value of $\mathbf{r}_{excluded}$ which corresponds to the allowed volume just touching the face of the tetrahedral cage, as determined from the lattice constant. It is seen that $\mathbf{r}_{excluded} > \mathbf{r}_{face}$ in the normal β phase, so that the Ag ions are confined to the central region of the tetrahedral site. As a result, the ionic conductivity is expected to be small and highly activated, as observed.[16,17] In the α phase, however, $\mathbf{r}_{excluded} < \mathbf{r}_{face}$ so that the Ag ions are allowed to flow easily through the multiply-connected tetrahedral regions. This is expected to yield a high conductivity with little activation energy, as observed.[16,17] It is interesting to note that the sharp transition from normal to superionic behavior is driven by the increase in $\mathbf{r}_{face}$ resulting from the HCP to BCC structural change in the iodine lattice, rather than by an unusual decrease in $\mathbf{r}_{excluded}$.

The resulting pair correlation function is asymmetric, as demanded by the data. The trends in $p_{Ag-I}(\mathbf{r})$ are shown in Fig. 4. At 77K it is a narrow Gaussian with $\sigma \approx 0.06$ Å. At room temperature in the β phase it has broadened and become somewhat asymmetric. At 198°C in the superionic α phase, the width and asymmetry is more pronounced, with a long tail extending to large near neighbor spacing. This corresponds to the Ag ion density flowing out of one tetrahedral location into another, and becoming large in the intervening trigonal sites.

It is interesting to note that the peaks in $p(\mathbf{r})$ which fit the EXAFS in all the above models are relatively narrow compared with the large Debye-Waller factors obtained from X-ray and neutron diffraction studies. The broadening in the nearest neighbor position which is observed in EXAFS, however, should not be confused with the much larger rms deviation from the average lattice position deduced from these studies $(\approx 0.3$ Å for iodine and 0.4 Å for Ag at $195°C$).[5] This broadening includes effects which do not broaden the EXAFS, such as long-wavelength acoustic phonons and certain types of static disorder.

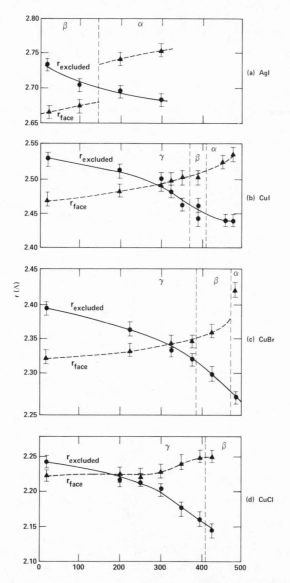

Fig. 3   The hard-sphere radius, $r_{excluded}$, obtained form a fit of the excluded volume model to the EXAFS data on (a) *AgI*, (b) *CuI*, (c) *CuBr* and (d) *CuCl*. Also included is r_{face}, the near neighbor cation-anion distance for a cation in the tetrahedral face. Note that $r_{excluded}$ becomes less than r_{face} when the ionic conductivity becomes large. Also note the differences in the variation of the parameters for *AgI* and the cuprous halides. The sharp change for *AgI* is reflected inthe sharp discontinuity in the ionic conductivity, whereas the smoother variation for the cuprous halide parameters is reflected in the smoother variation observed in their conductivity.

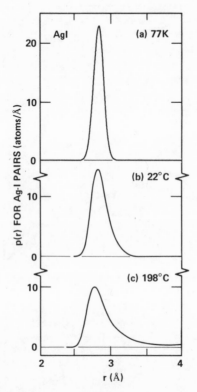

Fig. 4   The first neighbor pair correlation function for *AgI* at (a) 77K, (b) 22°C, and (c) 198°C. The narrow Gaussian of width 0.06 Å at 77K becomes broad and asymmetric at elevated temperature. This trend is described by the excluded volume model.

Discussion of Data on the Cuprous Halides

EXAFS measurements were performed on the Cu K-edge of $CuCl$, $CuBr$, and CuI as a function of temperature from 77K up into the superionic phases.[19,27,29] The 77K data served as the structural standard. At this temperature, all three materials have the cubic zincblende structure in which the Cu ions reside at the center of a halogen tetrahedron. In this case, the nearest neighbor pair correlation function, $p_{Cu-X}(\mathbf{r})$, is a narrow Gaussian with 4 halogen nearest neighbors centered at $\mathbf{r}_{tet}$. The width is estimated to be approximately 0.06 Å from our fit to the 77K data on $CuCl$ using calculated phase shifts and backscattering amplitude.[23] A width of 0.06 Å has also been obtained for $CuBr$ at 77K.[39] Four structural models were tested, as illustrated below using the EXAFS data on CuI at 470°C in the superionic α phase.

(1) The single Gaussian peak did not fit well. This is to be expected since the Cu ion vibrations cannot be described by a harmonic oscillator in this superionic phase. The resulting parameters are shown in Table 2.

(2) In the displaced site model, $p_{Cu-I}(\mathbf{r})$ is described by the sum of two Gaussians. Such an approach, proposed by Miyake et $al.$,[25] and Buhrer and Halg,[26] has the Cu ions in α-CuI statistically distributed over the four metastable sites that are displaced from the center of a tetrahedron toward the tetrahedral face. This class of models did not fit the data well, as seen from the resulting parameters in Table 2.

(3) The anharmonic oscillator model was proposed for α-CuI by Matsubara.[35] The potential in the γ and α phases where the iodine forms an FCC cage is taken to be of the form of Eq. 4 where a and b are temperature-dependent parameters. Matsubara obtained expressions for a and b from the diffraction and specific heat data of Miyake, et $al.$[25] With these specific parameters ($a_{Cu} = 1.8 \times 10^{-12}$ $erg/Å^2$ and $b_{Cu} = 2.9 \times 10^{-12}$ $erg/Å^3$ at 470°C), a good fit to the EXAFS data could not be obtained. For example, at 470°C this model yields $R=13.5\%$. A generalization of this anharmonic oscillator model, in which a and b are adjusted for a best fit, yielded $R=4.6\%$, still not a good fit to the data. The anharmonic model with terms up to third order does not fit the data because it is not sufficiently anharmonic. It does not provide a good approximation to the hard sphere repulsion, which is important for the superionic conductors since the ions approach closely to one another as the mobile ions move through the immobile ion cage. The excluded volume model treats this term as the dominant interaction and therefore yields a good fit to the data.

(4) For applications of the excluded volume model to FCC structures such as α-CuI, three parameters are needed. $\mathbf{r}_{excluded}$ and σ, as for BCC structures, and the concentration of Cu ions in the octahedral site, c_{oct}. The Cu ion distributions $within$ the tetrahedral and octahedral volumes are determined by the softened hard-sphere interaction. For α-CuI at 470°C, this yields $R=1.8\%$ for $\mathbf{r}_{excluded} = 2.44$ Å, $c_{oct} = 30\%$, and $\sigma = 0.08$ Å, as shown in Table 2. This R is significantly better than the values of $R >$ 3.5% obtained for the other models tested (as is the case at all temperatures studied). The parameters $\mathbf{r}_{excluded}$ and c_{oct} determined from the EXAFS data are shown in Fig. 3b and Fig. 5a, respectively. Also included in Fig. 3b is $\mathbf{r}_{face}$, the near neighbor $Cu-I$ spacing for a Cu ion at the center of the face shared by a tetrahedron and octahedron, obtained from the lattice constant. From these figures, it is seen that the concentration of the Cu ions in the octahedral sites begins to increase when $\mathbf{r}_{excluded}$ becomes less

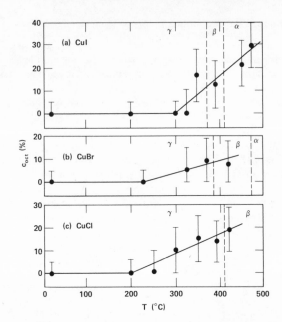

Fig. 5   The concentration of *Cu* ions in the octahedral sites versus temperature
            obtained from a fit of the excluded volume model to the EXAFS data of (a)
            CuI, (b) *CuBr*, and (c) *CuCl*. Note that c_{oct} becomes significant when the
            ionic conductivity becomes large. Also note the similarity in the variation of
            c_{oct} for all three materials. Since the structure is BCC for $\alpha-CuBr$, c_{oct} is
            not needed as a third parameter.

than r_{face}, allowing easy passage through that face. This is also the temperature region
where the ionic conductivity (see Fig. 1) becomes significant (*i.e.,* $\approx 10^{-3}$ (Ω cm)$^{-1}$).
These results are in agreement with the conduction path consisting of alternating
tetrahedral and octahedral locations, passing through the shared faces.[38] The resulting
nearest neighbor pair correlation functions are qualitatively similar to those for $Ag-I$
shown in Fig. 4. They differ principally in that the octahedral occupation enhances substantially the long tail which develops at large **r** as the conductivity increases.

   In order to determine the energy difference between the tetrahedral and octahedral
sites from the concentrations c_{tet} and c_{oct}, one must take into account the substantially
larger allowed volume η_{oct} in the octahedral site. Let the energy difference between
sites be $\Delta V = V_{oct} - V_{tet}$. Using $\rho = N_0 c/\eta$, where N_0 is the total number of *Cu* ions,
and the Boltzmann relation, one has

$$c_{oct}/c_{tet} = c_{oct}/(1-c_{oct}) = (4\eta_{oct}/8\eta_{tet})exp(-\Delta V/k_B T) \quad . \qquad (9)$$

At 470°C we obtain $c_{oct} \approx 0.3$ and $\eta_{oct}/\eta_{tet} \approx 10$, so that Eq. 9 yields $\Delta V \approx 0.16$ eV.
This is the energy barrier for *Cu* motion through the lattice and compares favorably
with the 0.2 eV activation energy determined from NMR relaxation measurements[40]
and ionic conductivity.[41]

The results on $CuBr$[29] parallel those on CuI and AgI, $\gamma-$ and $\beta-CuBr$ being similar to $\gamma-$ and $\beta-CuI$ and $\alpha-CuBr$ similar to $\alpha-AgI$. For $\gamma-$ and $\beta-CuCl$,[29] the results are similar to those on the γ and β phases of CuBr and CuI. The comparison of the various models for $CuBr$ and $CuCl$ is presented in Table 3, and the resulting parameters of the excluded volume model are shown in Figs. 3 and 5. The same relationships between the parameters and the ionic conductivity that were observed for CuI exist as well for $CuBr$ and $CuCl$.

Table 3. The parameters obtained for several structural models when adjusted to fit the data on CuBr and CuCl. The uncertainty in the determined distances is 0.01 Å and in the other parameters is $\approx 10\%$.

Material	Model	R (%)	N	r_0 (Å)	σ (Å)
$\gamma-$CuBr 370°C	Tet site		4	2.48	
	Oct site		6	2.87	
	Harmonic	4.2	4	2.40	0.15
		3.4	2.7	2.39	0.13
	Displaced site:	3.7	3	2.39	0.13
	$<111>$		1	3.13	0.31
	Anharmonic	3.0	(see Ref. 29)		
	Excluded volume	1.1	$[r_{excluded}=2.31; c_{oct}=9\%]$		
$\gamma-$CuCl 393°C	Tet site		4	2.36	
	Oct site		6	2.73	
	Harmonic	6.3	4	2.32	0.17
		5.6	2.88	2.31	0.14
	Displaced site:	3.7	3	2.30	0.14
	$<111>$		1	2.68	0.21
	Anharmonic	3.8	(see Ref. 29)		
	Excluded volume	0.6	$[r_{excluded}=2.16; c_{oct}=14\%]$		

Mobile Ion Density

The excluded volume model can also be used to obtain charge contour plots accord-
ing to Eq. 7. The parameters that specify the cation density $\rho_c(\mathbf{r})$ (i.e., $\mathbf{r}_{excluded}$, σ,
and c_{oct}) were determined in fits to the data as discussed above and are shown in Figs.
3 and 5. Various cross sections of $\rho_c(\mathbf{r})$ can then be obtained to vividly illustrate the
spreading of the mobile ion density with increasing temperature and to indicate the conduction path taken from one site to another.

Fig. 6 displays the cation density on the face of the BCC unit cell of superionic
$\alpha-AgI$ at 198°C. The plot is of the (100) plane where the anions occupy the corners
of the square and the positions above and below the center of the square. The Strock
sites[2] are labeled by symbols as noted. It is seen that the cation density peaks at the
tetrahedral sites but is also substantial at the bridging trigonal sites. The easy conduction path is along the $<110>$ directions from one tetrahedron to another through the
shared faces. The charge density at the octahedral site is lower than that for the trigonal site indicating that the $<100>$ direction is a less probable path for conduction.
One can also estimate the potential energy barrier heights for conduction in different
directions. The distorted tetrahedral site is the lowest energy site for Ag ions in this
iodine BCC cage. From Eq. 3, the trigonal site is higher in energy by about 0.04 eV
while the octahedral site is higher by about 0.18 eV. The 0.04 eV barrier for conduction along the $<110>$ direction is close to the 0.05 eV conductivity activation energy.[17]
For conduction along the $<100>$ direction, as suggested by Flygare and Huggins,[34] the
barrier is approximately 0.18 eV. This is nearly five times larger than the barrier along
the $<110>$ direction, indicating that conduction in the $<100>$ direction is substantially less favorable.

It is interesting to compare the Ag charge distribution shown in Fig. 6 with the qualitatively similar distribution which has been extracted from neutron diffraction measurements on single-crystal $\alpha-AgI$ at 160°C by Cava, et al. (see Fig. 1a in that work).[9]
The ratio of ρ at a trigonal site to that at a tetrahedral site is in agreement for the two
studies, being about 0.4. The octahedral densities differ, being $\rho/\rho_0 \approx 0.01$ from the
EXAFS and about 0.15 from the neutron data. A general qualitative difference is that
our results show a more abrupt decrease in the Ag ion density as an iodine ion is
approached. This reflects the essential differences between our softened hard-sphere
pair potential and the anharmonic potential well used by Cava et al., which incorporates
a much softer core-core repulsion.

Next we consider the cation density in the FCC materials. Fig. 7 displays the Cu
ion density in the (110) plane of the iodine FCC lattice at four temperatures for CuI.
The dramatic spreading of the Cu density with increasing temperature is clearly evident.
At room temperature (Fig. 7a), only the tetrahedral sites are occupied. At this temperature CuI has the zincblende structure, so only alternate tetrahedral sites are occupied (T_d^2). Since our EXAFS analysis treated only nearest neighbors, the difference
between alternate tetrahedral sites being occupied (T_d^2) and all the tetrahedral sites
being occupied (O_h^5) could not be distinguished. For this we rely on the results of
diffraction studies.[26] At 350°C (Fig. 7c) lobes of Cu ion density have entered the
octahedral site $(c_{oct} = 9\%)$ and are even more substantial in the superionic α phase at
470°C (Fig. 7d) where $c_{oct} = 30\%$. In the α phase the tetrahedral sites are equally occupied (O_h^5).[26] The tetrahedral site is the lowest energy site. Along the $<111>$ direction
from one tetrahedron to another through an octahedron, the octahedral site is a local

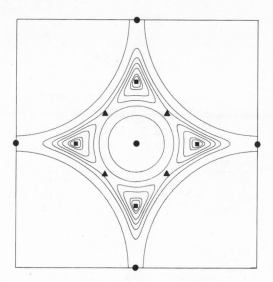

Fig. 6   A contour plot of the *Ag* ion density inthe superionic α phase of *AgI* at
            198°C. The plot is for a (100) plane which inclues the conduction path in
            the <110> directions from one tetrahedron through a trigonal face into
            another tetrahedron. The various Strock sites are labeled as follows: ■,
            tetrahedral; ●, octahedral; and ▲, bridging face site. Moving away from the
            tetrahedral centers, the contours correspond to $\rho/rho_0 =$ 0.95, 0.9, 0.7, 0.5,
            0.3, 0.1, $\approx$ 0. Note that the cation density peaks at the tetrahedral sites but
            also spreads substantially through the tetrahedral faces.

minimum in density. This difference in density yields a barrier height for conduction in
the <111> direction of approximately 0.16 ev. The density is very small at
(1/4,1/4,0), the edge of the tetrahedron shared with an octahedron. The activation
barrier for this site is approximately 0.7 eV, substantially larger than the barrier at the
tetrahedral face. This implies that the viable conduction path is in a <111> direction
through a face into an octahedron, not in a <100> direction through a tetrahedral
edge.

Ionic Conductivity

In the excluded volume model, the cations move as a gas in the allowed regions of
the anion lattice. Those allowed regions are centered at the tetrahedral sites where the
cations are confined by the hard-sphere anion walls. They remain in this cavity until
their motion is directed toward the opening in the tetrahedral face and they travel to a
neighboring region. Having obtained an expression for the three-dimensional potential
which governs this motion, it is now possible to calculate the DC conductivity.[42]

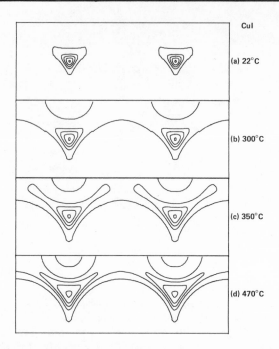

Fig. 7   A contour plot of the Cu ion density in the (110) plane of CuI in the γ
phase at (a) 22°C, (b) 300°C and (c) 350°C and in the superionic α phase at
(d) 470°C. The plot is for the (110) plane which includes a conduction path
in <111> direction from one tetrahedron through an octahedron, into
another tetrahedron. moving away fromthe tetrahedral sites, the contours
correspond to $\rho/\rho_0 =$ 0.95, 0.5, 0.2, 0.05,Å 0. Note that the Cu ion density
peaks at the tetrahedral sites but also spreads through the tetrahedral faces
into the octahedral regions.

To zeroth order, the cations are treated as a dilute solution of charged particles in a
Boltzmann gas of heavy molecules, for which the conductivity is

$$\sigma_{BG} = (3\pi^{1/2}/8)(n\ e^2\ L)(2\mu k_B T)^{-1/2}\ ,\qquad\qquad (10)$$

where n is the cation density, L is the effective scattering length, and μ is the reduced
mass. For AgI at 198°C, σ_{BG} is approximately 12 $(\Omega\ cm)^{-1}$ for reasonable values of L
and μ. This is eight times the observed[17] conductivity of 1.5 $(\Omega\ cm)^{-1}$.

Next, a potential which reflects the constraints of the anion lattice on the flow of the
cations is imposed on this Boltzmann gas. In this step, only the three-dimensional $V(\mathbf{r})$
obtained above, arising from the cation-anion core-core repulsion, is included and the
effects of possible correlations among the cations are neglected. The probability that an

ion at a point in the allowed volume r_i will actually escape from that volume along a path j is determined by the maximum potential, V_j, encountered along the path according to the usual Boltzmann factor, $P_{ij}(T) = exp[-(V_j - V(r_i))/k_B T]$, which is never in excess of unity. Sums are then performed along all the paths j associated with each allowed point, including the appropriate weighting factors. If the probability of a cation occupying r_i is w_i, then the total escape probability from a given allowed volume is

$$P(T) = \Sigma_i \; w_i \; \Sigma_j \; P_{ij}(T)/4\pi \; \Sigma_i \; w_i \; , \tag{11}$$

where the sum over j includes all escape paths from the point i and the sum over i includes all the allowed points. The final expression for the DC conductivity is $\sigma(T) = \sigma_{BG} P(T)$, determined completely by the parameters for the Boltzmann gas and the structural information obtained from the EXAFS experiments.

For $\alpha-AgI$ at 198°C, the escape probability $P(T)$ deduced from the EXAFS experiments is approximately 0.07. The corresponding value of $\sigma(T)$ is 0.8 $(\Omega \; cm)^{-1}$. This calculation has been repeated for AgI at four temperatures, and is shown in Fig. 8 together with the measured conductivity.[16,17] Note that the predicted conductivity has been increased by a factor of approximately two to bring it into agreement with experiment at high temperatures, a reasonable procedure given the uncertainties in the parameters which determine σ_{BG}. Considering for the moment just the high temperature superionic α phase, it is clear that the predicted slope is quite good. That is, the constraints of the immobile ion lattice as incorporated in $V(r)$ explain very well the observed[17] "activation energy" of 0.05 eV.

In considering the low temperature β phase, there is an additional complication. In the α phase of AgI, the volume considered in evaluating Eq. 11 included all of the relevant volume in the solid, even the so-called trigonal and octahedral sites. In the wurtzite β phase, the rest site is similar, having tetrahedral symmetry. The conduction path, however, must now include octahedral sites. Since these sites are known to be unoccupied from diffraction studies, we can obtain no information about the potential near them from the structural studies. Accordingly, we introduce an arbitrary site potential energy difference ΔV into the calculation of V_j to obtain the escape probability. Since we know from diffraction studies that ΔV cannot be zero, it is no surprise that the top curve in Fig. 8 does not fit the measured conductivity in the β phase. From the comparisons in Fig. 8, we deduce that the appropriate ΔV is between 0.5 and 0.6 eV at low temperatures.

It is interesting to note that the measured conductivity curves upward as the transition to superionic behavior is approached from below. This behavior is characteristic of these materials.[16-18] We attribute this effect to a decline in ΔV as the transition is approached from below, one aspect of the incipient melting of the cation lattice.[15]

We have also used this model to calculate the ionic conductivity for the Cu halides, with comparable agreement. As in the case of AgI, ΔV becomes roughly 0.5 eV in the normal low temperature phases.

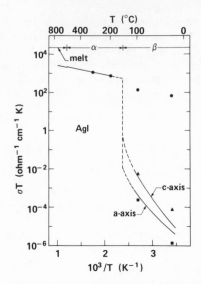

Fig. 8   The predicted conductivity for *AgI* is shown for activation energies ΔV of 0, 0.5, and 0.6 eV (solid curves, triangles, and squares, respectively). The measured conductivity is also shown in the β and α phases (see References 16 and 17).

Summary

The EXAFS data on *AgI* and the cuprous halides from low temperatures up into the superionic phases have been analyzed using four structural models: harmonic oscillator, displaced site, anharmonic oscillator, and excluded volume. The most satisfactory description of the EXAFS structural information is obtained with the excluded volume model. In this model the actual ionic pair potential is approximated by a softened hard-sphere interaction which, in turn, determines the near neighbor pair correlation function. The results indicate that the tetrahedral locations in the halogen lattice are preferred by the mobile cations but that substantial density occurs at the trigonal sites at elevated temperatures. For the BCC materials, $\alpha-AgI$ and $\alpha-CuBr$, this indicates that the mobile ions move in <110> directions through the face-shared tetrahedra. For the FCC materials, $\alpha-CuI$ and $\gamma-$ cuprous halides, significant octahedral occupation was also found at elevated temperatures. The indicated conduction path in these cases is in the <111> directions through the faces shared by the tetrahedra and octahedra in the FCC halogen lattice.

The structural results also yield some insight into the transition to the superionic phase. For *AgI* where the ionic conductivity increases sharply by four orders of magnitude at the β to α transition, there is an abrupt opening of the tetrahedral face allowing the *Ag* ions to move into neighboring locations. For the cuprous halides, on the other hand, where the conductivity increases more gradually, there is a more gradual change in the structural parameters.

In addition to determining the conduction direction from the structural results, one can estimate the potential energy barrier heights and calculate the temperature-dependent DC ionic conductivity. For example, in $\alpha-AgI$ the trigonal site presents a barrier of about 0.04 eV and the octahedral site a barrier of about 0.18 eV. Conduction in <110> directions through the trigonal site is preferred, leading to an activation energy of approximately 0.04 eV in good agreement with conductivity studies. In $\alpha-CuI$ the octahedral site presents a barrier of about 0.16 eV and the tetrahedral edge a barrier of 0.7 eV. Conduction in the <111> directions through the octahedra is preferred, leading to an activation energy of 0.16 eV in agreement with conductivity and NMR studies. The conductivity is calculated by treating the conducting ions as a Boltzmann gas in the presence of the three-dimensional potential determined by fitting the pair distribution functions obtained from EXAFS data. The resulting temperature-dependent DC conductivities fit the experiment on *AgI* and the cuprous halides. By working equally well in both the normal and superionic phases, this model supports strongly the concept that the conduction mechanism is essentially unchanged on the transition to superionic behavior.

ACKNOWLEDGEMENTS

We wish to acknowledge the contributions of J. C. Mikkelsen, Jr., and W. Stutius. Some of the materials incorporated in this work were developed at the Stanford Synchrotron Radiation Laboratory which is supported by the National Science Foundation (under Contract DMR77-27489), in cooperation with SLAC and the Department of Energy.

REFERENCES

1. Several review articles have appeared recently: (a) *Solid Electrolytes,* ed. S. Geller, Vol. 21 of Topics in Applied Physics, Springer-Verlag, Berlin (1977); (b) *Solid Electrolytes: General Principles, Characterization, Materials, Applications,* ed. P. Hagenmuller and W. von Gool, Academic Press, New York (1977); (c) J. B. Boyce and B. A. Huberman, "Superionic Conductors: Transitions, Structures, Dynamics," Phys. Reports **51,** 189 (1979); (d) *Physics of Superionic Conductors,* ed. M. B. Salamon, Vol 15 of Topics in Current Physics, Springer-Verlag, Berlin (1979); (e) *Fast Ion Transport in solids,* ed. P. Vashishta, J. N. Mundy, and G. K. Shenoy, North-Holland, New York (1979).

2. L. W. Strock, Z. Phys. Chem. Abt. B **25,** 411 (1934); **31,** 132 (1936).

3. S. Hoshino, J. Phys. Soc. Japan **12,** 315 (1957).

4. G. Burley, J. Chem. Phys. **38,** 2807 (1963).

5. W. Buhrer and W. Halg, Helv. Phys. Acta **47,** 27 (1974).

6. A. F. Wright and B. E. F. Fender, J. Phys. C **10**, 2261 (1977).

7. G. Eckold, K. Funke, J. Kalus, and R. E. Lechner, J. Phys. Chem. Solids **37**, 1097 (1976), and references contained therein.

8. J. B. Boyce, T. M. Hayes, W. Stutius, and J. C. Mikkelsen, Jr., Phys. Rev. Lett. **38**, 1362 (1977).

9. R. J. Cava, F. Reidinger, and B. J. Wuensch, Solid State Comm. **24**, 411 (1977).

10. M. Suzuki and H. Okazaki, Phys. Stat. Sol. (a) **42**, 133 (1977).

11. S. Hoshino, T. Sakuma, and Y. Fujii, Solid State Comm. **22**, 763 (1977).

12. J. B. Boyce and J. C. Mikkelsen, Jr., Solid State Comm. **31**, 741 (1979).

13. For a discussion of the various types of transitions and a list of materials in each category, see Ref. 1c.

14. M. O'Keeffe and B. G. Hyde, Phil. Mag. **33**, 219 (1976) and references therein.

15. B. A. Huberman, Phys. Rev. Lett. **32**, 1000 (1974).

16. H. Hoshino and M. Shimoji, J. Phys. Chem. Solids **35**, 321 (1974).

17. A. Kvist and A. M. Josefson, Z. Naturforsch **23A**, 625 (1968); H. Hoshino, S. Makino and M. Shimoji, J. Phys. Chem. Solids **35**, 667 (1974); R. N. Schock and E. Hinze, J. Phys. Chem. Solids **36**, 713 (1975); P. C. Allen and D. Lazarus, Phys. Rev. B **17**, 1913 (1978).

18. J. B. Wagner and C. Wagner, J. Chem. Phys. **26**, 1597 (1957) and references contained therein; T. Jow and J. B. Wagner, J. Electrochem. Soc. **125**, 613 (1978).

19. For a discussion of EXAFS and applications to superionic conductors see J. B. Boyce and T. M. Hayes, "Structure and Its Influence on Superionic Conduction: EXAFS Studies", Chapter 2 of Ref. 1d.

20. For a discussion of EXAFS with references to earlier work see the review by T. M. Hayes, J. Non-Cryst. Solids **31**, 57 (1978).

21. N. F. Mott and H. S. W. Massey, *The Theory of Atomic Collisions,* third ed., Clarendon Press, Oxford, 1965, p. 562.

22. T. M. Hayes, P. N. Sen, and S. H. Hunter, J. Phys. C **9**, 4357 (1976).

23. B.-K. Teo and P. A. Lee, J. Am. Chem. Soc. **101**, 2815 (1979).

24. P. H. Citrin, P. Eisenberger and B. M. Kincaid, Phys. Rev. Lett. **36**, 1346 (1976).

25. S. Miyake, S. Hoshino, and T. Takenaka, J. Phys. Soc. Japan **7**, 19 (1952).

26. W. Buhrer and W. Halg, Electrochim. Acta **22**, 701 (1977).

27. J. B. Boyce, T. M. Hayes, J. C. Mikkelsen, Jr., and W. Stutius, Solid State Comm. **33**, 183 (1980).

28. S. Hoshino, J. Phys. Soc. Japan **7**, 560 (1952).

29. J. B. Boyce, T. M. Hayes, and J. C. Mikkelsen, Jr., Solid State Comm. (in press).

30. M. Sakata, S. Hoshino and J. Harada, Acta Cryst. A **30**, 655 (1974).

31. J. Schreurs, M. H. Mueller, and L. H. Schwartz, Acta Cryst. A **32**, 618 (1976).

32. T. M. Hayes, J. B. Boyce, and J. L. Beeby, J. Phys. C **11**, 2931 (1978).

33. For a review of the molecular dynamics work, see the papers in Ref. 1e by A. Rahman, p. 643; W. Schommers, p. 625; and P. Vashishta and A. Rahman, p. 527.

34. W. H. Flygare and R. A. Huggins, J. Phys. Chem. Solids **34**, 1199 (1973).

35. T. Matsubara, J. Phys. Soc. Japan **38**, 1076 (1975).

36. J. Harada, H. Suzuki, and S. Hoshino, J. Phys. Soc. Japan **41**, 1707 (1976).

37. V. Valvoda and J. Jecny, Phys. Stat. Sol. (a) **45**, 269 (1978).

38. L. V. Azaroff, J. Appl. Phys. **32**, 1658 (1961).

39. B. Bunker, private communication.

40. J. B. Boyce and B. A. Huberman, Solid State Comm. **21**, 31 (1977).

41. W. Jost, *Diffusion in Solids, Liquids and Gases,* Academic Press, New York, 1960, p. 188.

42. T. M. Hayes and J. B. Boyce, Phys. Rev. B **21**, 2513 (1980).

EXTENDED X-RAY ABSORPTION FINE STRUCTURE STUDIES

AT HIGH PRESSURE

R. Ingalls, J.M. Tranquada and J. E. Whitmore

Department of Physics
University of Washington
Seattle, WA 98195

E.D. Crozier and A. J. Seary

Department of Physics
Simon Fraser University
Burnaby, B.C. V5A 1S6

Introduction

It is the purpose of this paper to review the problems and possibilities offered by EXAFS studies of materials at high pressures. This is a new field for which high pressure techniques must be refined. Cell designs and methods of measuring the pressure in situ which we have used are indicated. It will be seen that we are currently concerned with the accuracy in measuring bond lengths as a function of pressure and, learning what EXAFS can tell us about pressure-induced structural or electronic phase changes.

Experimental

The method we have chosen consists of introducing a layer of sample into a gasket that is squeezed between two anvils of various materials. In our initial work[1] the radiation passed through a boron and lithium hydride gasket in a direction perpendicular to the axis of the tungsten carbide anvils, (Fig. 1a). More recently[2] we have used "X-ray-transparent" materials such as boron carbide or diamond as tips on hollowed out anvils, with the radiation along the axis (Fig. 1b). In either case the sample area was $\sim 10^{-1} mm^2$ with the radiation passing through $\sim$3mm of relatively low Z material. We are therefore obviously at a distinct disadvantage when studying low-energy absorption edges. The increased photon flux available at SSRL since the installation of focussing mirrors and wiggler magnets has greatly enhanced our capabilities. Pressure studies are now feasible over the energy range 6kev to 25kev.

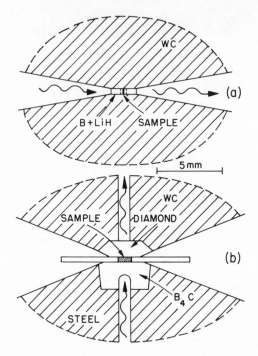

Fig. 1.   Experimental high pressure cell geometry: (a) Radiation perpendicular to
the load axis; (b) Radiation parallel to the load axis.

Review of Previous Work

Several years ago we used the focussed EXAFS beam line at SSRL to study the iron
K-edges in FeF_2 and FeS_2.[1] The data was analyzed by simply transforming the $\chi(k)$
data and using the peak in the magnitude of the transform $|\phi(R)|$ to locate neighbor-
ing atom positions. The $Fe-Fe$ distances were observed to contract at the same rate as
previous pressure-volume work had shown. However, this simplified analysis seemed
to indicate the $Fe-S$ bond in FeS_2 was pressure independent. Very recently we have
refined this analysis by the Fourier filtering technique[3] and observed the $Fe-S$ bond
length, in fact, to scale with the $Fe-Fe$ separation. This appears to be consistent with
recent X-ray diffraction studies.[4]

Fourier filtering techniques were also employed to determine the bond compressibilities in *NaBr* and *Ge*.[2] Precise pressure determinations were made employing the shift of the ruby fluorescence line in conjunction with a diamond anvil. However, because of Bragg peaks from the diamond interfering with the EXAFS signal, (Fig. 2) the accuracy with which bond length changes could be determined was only about ± 0.01Å; the method should be capable of an accuracy of' $\sim \pm 0.003$Å. However, the pressure-induced changes in the bond lengths determined via EXAFS, to an upper pressure of 21kbar in *NaBr* and 40kbar in *Ge*, agreed with the changes observed in X-ray diffraction and pressure-volume data. Thus, at least to the upper pressure limits currently investigated, the pressure-induced changes in the EXAFS phase shifts appear to be very small.

This work[2] was also useful in developing *NaBr* as an EXAFS pressure standard, to replace the diamond-anvil, ruby fluorescence technique. In fact, we soon hope to be able to use routinely the EXAFS of *NaBr*, and then *RbCl*, as our pressure gauges.

The compound *RbCl* is, of course, interesting because of its change of phase from the $B1$ (*NaCl*) to the $B2$ (*CsCl*) structure at 5.2kbar. In a preliminary report[5] we showed how the near edge signature of the *Rb* K-edge changes markedly as the compound transforms. The same is true of both K-edges in *CuBr* which undergoes a cubic to tetragonal transformation at $\sim$45kbar.[6]

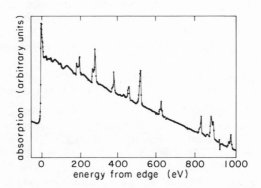

Fig. 2.   Spectrum of the K-edge of *Ge* showing Braff peaks caused by the diamond anvil.

Recent Results

Alkali halides. We are currently embarked upon a program to study the EXAFS of alkali halides as they transform from the $B1$ to $B2$ structure. Simultaneously we wish to further our accuracy in measuring their pressure response so that they may be developed as pressure standards.

In Fig. 3 we show the effect of pressure on the *Rb* K-edge EXAFS peaks of the $B2$ form of *RbCl*. These peaks shift with pressure in much the same manner as in *NaBr*.[2] Both materials have been run simultaneously as a cross check. We have also studied the changes of lattice spacing in these materials when run simultaneously. Such an analysis is still in progress, but preliminary results seem to indicate a rather precise correlation with that expected from pressure volume studies[6] (Fig. 4). We hope to extend this type of study to much higher pressures. The material *NaBr* is expected to transform to the $B2$ structure somewhat above 300kbar, whereas *RbCl* is expected to remain in that phase at all pressures above 5.2kbar.

We have also begun an investigation of the similar behavior in *KBr* which undergoes the $B1-B2$ transformation at 17.2kbar. Fig. 5a shows the near edge structure of *NaBr*, *RbCl*, and *KBr* for the $B1$ phase, while Fig. 5b shows a comparison between the spectra of *RbCl* and *KBr* in the $B2$ phase. The reasons for the change in the spectra before and after the transition await a detailed explanation.

Further analysis will include a study of the EXAFS-determined disorder term (Debye-Waller factor) as a continuous function of pressure and its change at the $B1-B2$ transition. The decrease of the disorder term with pressure has been documented earlier for FeF_2 and FeS_2^1 as well as *NaBr*.[2] It is also quite apparent in Fig. 3 for *RbCl*, manifesting itself as an enhancement of the EXAFS oscillation with pressure.

Cuprous bromide. We have previously reported[5] the changes of the near edge structure, for both K-edges, as this material transforms from the cubic, zinc blende phase *CuBrIII* to tetragonal *CuBrV* at approximately 45kar. Further analysis has indicated that the near neighbor $Cu-Br$ bond length, as well as the associated disorder term, go through a discontinuity at the transition, Fig. 6. Even more interesting is the fact that the EXAFS results for this bond length do not indicate the expected contraction.[6] It is not known to what extent this is due to a pressure-dependent phase shift perhaps brought about by charge transfer. We believe that anharmonic motion of the *Cu* atoms[6] may be a partial explanation. It has been shown previously that such an effect can interfere with a determination of thermal expansion of bond lengths.[9,11]

It is pertinent to note that the detailed structure of *CuBrV* has not been determined yet. From our preliminary results using the magnitude of the Fourier transform, Fig. 7, the largest change occurs in the second nearest neighbor bond length. The coordination number of nearest neighbors does not appear to change. We intend to pursue the study of this material and extend it to higher pressures which will include another structural phase transition, tetragonal to $B1$ (*CuBrV* to *CuBrVI*) at approximately 65kbar.

Samarium selenide. This material has the *NaCl* structure and undergoes under pressure a continuous structureless transition from the insulating state to the "mixed valence" conducting state.[10] At atmospheric pressure the *Sm* is the divalent or $4f^6$

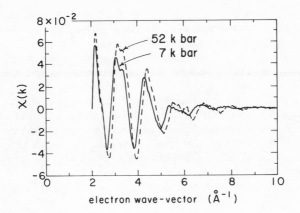

Fig. 3. The EXAFS interference function $\chi(k)$ of the *Rb* K-edge in *RbCl*.

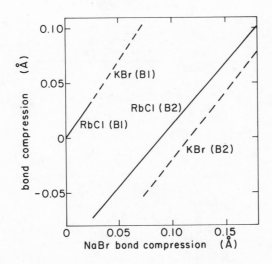

Fig. 4. Expected compression of the bond lengths in *RbCl* and *KBr* as a function of that in *NaBr*, from data of Ref. 6.

configuration, whereas under pressure it approaches the $4f^5 5d$ state, which, because of the delocalized outer $5d$ electron, also approaches the trivalent state.

We have studied the *Sm* L-edges and the *Se* K-edge in this compound. Fig. 8 illustrates what we observe for the near edge structures. The pronounced change observed for the L3 edge is tentatively interpreted as a change from the predominantly $4f^6$ configuration to the $4f^5$, the absorption edge of which lies at higher energies. Coulomb effects would quite reasonably account for such behavior because the $2p$ electron is expected to be more tightly bound in the $4f^5$ configuration. The $L2$ edge is virtually identical to the $L3$ edge both in appearance and behavior. The $L1$ edge also appears to shift towards higher energy with pressure, but it lacks the "white line" feature. It should be remarked that such good *Sm* L-edge spectra were made possible by the new wiggler at SSRL. We still do not have proper statistics for an EXAFS analysis of such low energy edges. The situation should further improve with the new focussing capability on the beam line.

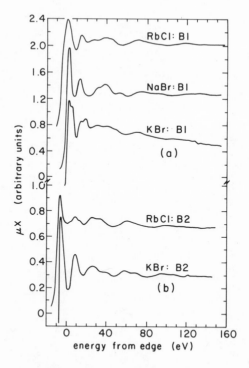

Fig. 5.   *Br* and *Rb* K-edges in (a) the B1 phases of *NaBr*, *RbCl*, and *KBr*, and (b) the B2 phases of *RbCl* and *KBr*.

We also show the effect of pressure on the *Se* K-edge (Fig. 8b). The broadening of the white line presumably can be accounted for when one calculates the electron configuration associated with the additional charge deposited onto the *Se* site with pressure. The EXAFS from this edge are of very high quality and clearly indicate the great contraction of the material as it undergoes compression, together with a partial valence change. This work is continuing.

Solid and liquid gallium. Pressure-dependent structural studies constitute a relatively unexplored area of liquid metal physics. Liquid metals can be classified into two categories, those for which the structure factor $S(q)$ at atmospheric pressure can be described to a first approximation by hard sphere models and those which cannot. For metals in the first category, such as the alkalis, theoretical understanding exists for the effective ion-ion interaction potential and the resulting liquid structure. In the second

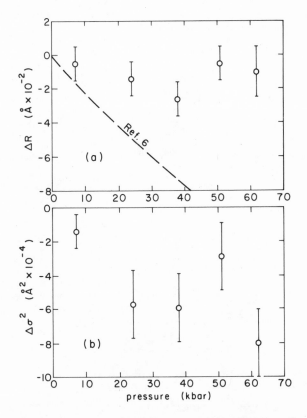

Fig. 6.   (a) Bond length and (b) its mean square deviation for the $Cu-Br$ nearest neighbor distance, as determined by *Br* K-edge EXAFS. The pressure scale is preliminary and may be subject to slight modification.

category are metals such as *Ga* which display a shoulder on the high q side of the first peak in $S(q)$. The origin of this shoulder and the local liquid structure are not fully understood.[12,13]

$S(q)$ has been measured as a function of pressure in X-ray[14-15] and neutron[16] diffraction experiments. The measurements, which for the most part have been at pressures $\leq$ 6kbar, have been restricted to the most compressible metals, the alkalis and, the momentum transfer range $1\text{Å}^{-1} \leq q \leq 4\text{Å}^{-1}$. The data were used to obtain information regarding the density-dependence of the ion-ion interaction potential and, the validity of various statistical mechanical models of the liquid state.

Our pressure-dependent EXAFS work differs in two respects from the diffraction studies. First, with our experimental method, pressures an order of magnitude higher are readily achieved and thus the study of metals other than alkalis is feasible. Secondly, the structural analysis of the EXAFS interference function $\chi(k)$ for liquid metals typically covers the momentum range $3\text{Å}^{-1} \leq k \leq 12\text{Å}^{-1}$ (or the momentum transfer range $6\text{Å}^{-1} \leq q \leq 24\text{Å}^{-1}$). As indicated elsewhere,[11] the high k values available with EXAFS data can be used to advantage in specifying the repulsive portion of the ion-ion interaction potential of liquid metals. It is this small R region of the potential and its pressure dependence which are of particular interest in determining the structure of liquid *Ga*.

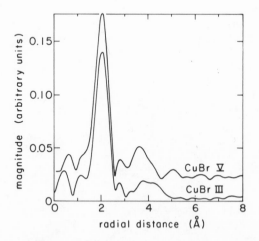

Fig. 7.   The magnitude of the Fourier transform obtained from *Br* K-edge EXAFS of
CuBr.

The first part of the *Ga* study involved the determination of reference EXAFS parameters. These were obtained by orienting a single crystal of α-*Ga* (orthorhombic unit cell with 1 neighbor at 2.47Å and 2 atoms at each of the following distances 2.70, 2.74 and 2.79Å) along the shortest bond direction. The phase shift $\delta(R)$ and the disorder term σ_1^2 were extracted by using the backscattering amplitude $T(k)$ from Teo and Lee[17] corrected for multielectron effects. The magnitude of σ_1^2 at 20°C was found to be $\sim.005Å^2$ which is more characteristic of a covalent bond such as occurs in $Ge\,(\sigma_1^2 \sim .0035Å^2)$[11] than a metallic bond such as occurs in $Zn\;(\sigma_1^2 \sim .009Å^2)$.[11] The remaining nearest neighbors of *Ga* were found to have a much larger σ^2 and a different $\delta(k)$ implying that asymmetry effects could not be neglected. In a study of polycrystalline *Ga* at 1 bar and 20°C the Fourier transform of $\chi(k)$ was dominated by the 6 neighbors located near 2.74Å although the shorter bond at 2.47Å was clearly evident as a strong beat frequency in the Fourier filtered $\chi(k)$. Supercooled liquid Ga under the same conditions did not show unambiguous evidence of beating. However, the distance to the main R-space peak was 0.2Å shorter than in polycrystalline *Ga*. Whether

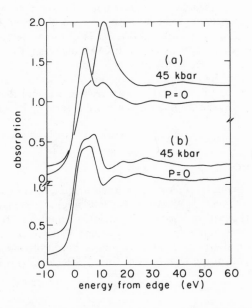

Fig. 8.   Near edge structure of *SmSe*; (a) *Sm* L3-edge and (b) *Se* K-edge.

this indicates a persistence in the liquid state of the local structure characteristic of the 2.47Å bond, a contraction due to asymmetry effects[9,11], or a combination is not yet clear.

Pressure-dependent measurements were made at 20°C spanning the α-phase to liquid and liquid to β-phase transitions. Significant changes were observed in R_1 and σ^2 relative to the supercooled liquid at 1 bar. The liquid will be studied further at higher temperatures and at higher pressures.

Conclusions

In this paper we have reviewed the experimental developments in high pressure X-ray absorption measurements. Severe constraints are imposed by the high pressure apparatus: thick cell walls and a sample cross-sectional area which is about two orders of magnitude smaller than used normally in EXAFS measurements. However, with the high photon flux available at the Stanford Synchrotron Radiation Laboratory high resolution X-ray absorption data ($\sim$.5ev) have been obtained. High pressure work, particularly the investigation of phase transitions, requires knowledge of the sample pressure while the experiments are being conducted. Unfortunately the conventional pressure determination from the shift in the ruby fluorescence line is precluded because it requires diamond anvils whose Bragg peaks will interfere, in general, with the EXAFS signal. Thus we have developed $NaBr$ and $RbCl$ as pressure calibrants that are included with the sample. Characteristic changes in the near edge and the extended fine structure permit on site pressure determinations.

The potential of X-ray absorption measurements for the investigation of pressure-induced structural or electronic phase changes has been illustrated by reference to the specific systems FeF_2, FeS_2, Ge, $NaBr$, KBr, $RbCl$, $CuBr$, $SmSe$, and Ga. One concern, a possible pressure dependence of the EXAFS phase shift $\delta(k)$, appears to have been negligible in the systems investigated. Consequently it is anticipated that pressure-induced changes in the nearest neighbor distance R_1 can be measured to ± 0.003Å. It is interesting to note that with increasing pressure the data analysis actually becomes easier and R_1 can be determined more accurately because of a decrease in the disorder term.

ACKNOWLEDGEMENTS

We are pleased to thank the following for their gracious assistance and communications: E. A. Stern, B. Bunker, J. W. Allen, F. Holtzberg, and A. Jayaraman. We were supported in part by the National Science Foundation (grant no. DMR 78-24995) and by the National Sciences and Engineering Research Council of Canada. The X-ray absorption measurements were made at the Stanford Synchrotron Radiation Laboratory with the financial support of the NSF (contract no. DMR 77-27489) in cooperation with the DOE.

REFERENCES

1. R. Ingalls, G. A. Garcia, and E. A. Stern, Phys. Rev. Lett. **40**, 334 (1978).

2. R. Ingalls, E. D. Crozier, J. E. Whitmore, A. J. Seary, and J. M. Tranquada, J. Appl. Phys., to be published.

3. F. W. Lytle, D. E. Sayers, and E. A. Stern, Phys. Rev. **B11**, 4825 (1975); E. A. Stern, D. E. Sayers, and F. W. Lytle, ibid, **B11**, 4836 (1975).

4. S. Yamaoka, O. Shimomura, H. Nakazawa, and O. Kukunaga, Proc. 7th Int. Conf. High Pres. (AIRPT), Pergamon, to be published.

5. R. Ingalls, J. E. Whitmore, J. M. Tranquada, and E. D. Crozier, Proc. 7th Int. Conf. High Pres. (AIRPT), Pergamon, to be published.

6. S. N. Vaidya and G. C. Kennedy, J. Phys. Chem. Solids **32**, 951 (1971).

7. V. Meisalo and M. Kalliomäki, High Temp. Pres. **5**, 663 (1973).

8. B. Bunker, private communication.

9. P. Eisenberger and G. S. Brown, Solid State Comm. **29**, 481 (1979).

10. A. Jayaraman, Proc. 6th Int. Conf. High Pres. (AIRPT), ed. K. D. Timmerhaus and M. S. Barber, Plenum, N.Y. (1979).

11. E. D. Crozier and A. J. Seary, Can. J. Phys. **58**, 1388 (1980); E. D. Crozier, elsewhere in this book.

12. D. Levesque and J. J. Weis, Phys. Lett. **60A**, 473 (1977); M. Silbert and W. H. Young, Phys. Lett. **58A**, 469 (1976).

13. K. K. Mon, N. W. Ashcroft and G. V. Chester, Phys. Rev. **B19**, 5103 (1979).

14. K. H. Brown and J. D. Barnett, J. Chem. Phys. **57**, 2009 (1972), ibid **57**, 2016 (1972).

15. K. Tsuji, H. Endo, S. Minomura, Phil. Mag. **31**, 441 (1975).

16. P. A. Egelstaff, D. I. Page, C. R. T. Heard, J. Phys. **C 4**, 1453 (1971); P. A. Egelstaff and S. S. Wang, Can. J. Phys. **50**, 2462 (1972), ibid **50**, 684 (1972).

17. B. K. Teo and P. A. Lee J. Am. Chem. Soc. **101**, 2815 (1979).

STRUCTURAL EVIDENCE FOR SOLUTIONS FROM EXAFS MEASUREMENTS

Donald R. Sandstrom

Department of Physics
Washington State University
Pullman, Washington 99164

B. Ray Stults

Corporate Research and Development Staff
Monsanto Company
St. Louis, Missouri 63166

R. B. Greegor

Boeing Company
Seattle, Washington 98124

Introduction

Physical evidence for the structure of solutions has been obtained by x-ray scattering,[1,2,3,4] neutron scattering,[5,6,7,8] Raman scattering,[9,10] transport measurements,[11] and NMR[12] techniques. In addition, extended x-ray absorption fine structure spectroscopy (EXAFS spectroscopy) has been applied to this problem. It is the sensitivity to local structure that makes EXAFS especially suitable for systems like this, for which no long range order is expected. Also, the element specificity of EXAFS means that the radial distribution functions deduced from EXAFS analysis contain only the relationship between atoms of the x-ray absorbing element and its neighbors. In contrast, neutron and x-ray scattering methods result in an average correlation function for the sample as a whole, unless special techniques such as isotopic substitution in the scattering of neutrons[6] or anomalous scattering of x-rays[13] are employed to distinguish the correlations between specific pairs of elements.

EXAFS has been applied to aqueous solutions by Eisenberger and Kincaid,[14] Sandstrom et al.,[15] Fontaine et al.,[16] Morrison et al.,[17] Sandstrom,[18] Huang et al.,[19] and Lagarde et al.[20] A brief discussion of these applications was given in a review by

Sandstrom and Lytle.[21] A more detailed discussion is given below. A somewhat specialized application of the general principles of solution EXAFS is the study of homogeneous catalysts.[22] A discussion of one such study is also presented.

Experimental Procedures

The majority of EXAFS studies of this type have been carried out using x-rays produced by a synchrotron radiation source. An exception is the work of Morrison et at.[17] in which measurements were made using a system based on a conventional x-ray generator.[23] The advantages of synchrotron radiation sources have been discussed by Doniach et al.[24] and by Winick and Bienenstock.[13] These include high intensity distributed smoothly over a broad spectral range, in addition to desirable polarization, collimation and time structure properties.

The design of sample cells depends somewhat on the type of detection technique to be used, i.e. whether the traditional absorption measurement is made, or whether the absorption is deduced by measurement of fluorescence x-ray emission.[25] In absorption measurements at low concentrations, the solvent absorption becomes the dominant component of the overall x-ray absorption. This is especially serious for lower Z solute atoms. In practice the x-ray path length in the sample must be adjusted for sufficient transmitted intensity for precise measurement, combined with a maximum absorption edge jump for the element studied. For a measurement of transmitted intensity limited by Poisson statistics, Jaklevic et al.[25] have shown that these considerations lead to an optimum sample thickness of $2/\mu_t$, where μ_t is the total linear absorption coefficient of the sample. For the case of dilute solutions, for which μ_t is essentially the same as the absorption coefficient of the solvent, Fig. 1 shows the variation of $2/\mu_t$ vs. x-ray photon energy for common solvents. Note that $2/\mu_t$ changes by almost three decades for x-ray energies between 2 and 20 KeV.

The measurements described by Sandstrom et al.[15,18] were carried out with samples confined in a mylar pouch sealed with silicone cement. A plexiglas frame provided thickness adjustment by means of shims, as shown in Fig. 2. Huang et al.[19] used a continuously adjustable cell[26] in an x-ray absorption edge study involving potassium solutions.

EXAFS studies of homogeneous catalysts under simulated reaction conditions were conducted using a special glass cell with a 5.0 cm path length between mylar windows. Rhodium K-absorption edge data were recorded in the absorption mode. The fixed path length was chosen as a best compromise for the concentration of the rhodium species of interest and the alcohol solvent. The reaction cell was equipped with outlets for de-gassing solvents and maintaining an inert atmosphere during the reaction. A septum was provided to allow introduction of reaction substrates to the reaction mixture. A magnetic stirrer was used to ensure reaction homogeneity.

In the absorption experiment, the fractional uncertainty of measurement of μ_x, the linear absorption coefficient of the solute element, scales as $1/C_x$, where C_x is the concentration of element x. In fluorescence detection, however, the scaling goes as $(1/C_x)^{1/2}$, giving this technique an advantage at lowest concentrations.[25] For these measurements the primary x-ray beam is incident on the same side of the sample from which the fluorescence x-rays are detected, and the sample should have a thickness of several absorption lengths $1/\mu_t$. Figure 3 illustrates an absorption spectrum measured by the fluorescence technique using a detector described by Stern and Heald.[27]

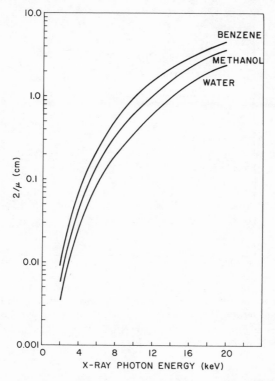

Fig. 1.   Optimum sample thickness $2/\mu$ vs x-ray photon energy, where μ is the linear absorption coefficient of the solvent alone, for three typical solvents.

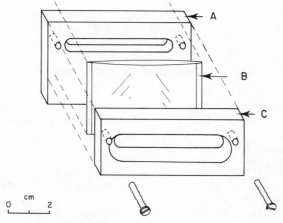

Fig. 2.   Adjustable liquid sample cell for absorption measurements. A Mylar pouch B is attached by double sided adhesive Mylar tape (not shown) to frames A and C. The spacing of the faces of the pouch is fixed by shims clamped between the frames.

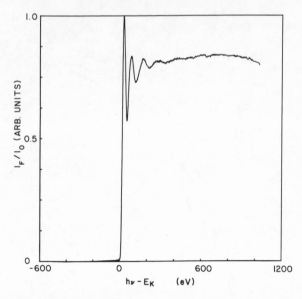

Fig. 3. Iron K-edge absorption spectra for a 0.03 mol/l ferric nitrate solution meas-
         ured by fluorescence technique. $E_k = 7112.0$ eV. The count rate in the
         fluorescence channel is $\approx$ 200 counts/s below the edge and $\approx$ 20,000 counts/s
         at the peak immediately above the edge.

Analysis for Structural Information

Analysis procedures have been extensively reviewed elsewhere,[21,28,29] and the
approach is substantially the same for solution systems. The objective is to obtain
N_i, r_i, σ_i^2, where these are the number of neighbors in the ith coordination shell, distance to neighbors in the ith shell, and the mean squared relative displacement of atoms
in the ith shell about distance r_i, respectively. In addition, chemical identification of
the ith shell neighbors can be made.

Because determination of the above structural parameters depends on knowledge of
the dependence of photoelectron wave-vector k of the phase function $\phi_i(k)$ and backscattering amplitude $F_i(k)$ for each shell, the present discussion will focus on determination of these quantities for solutions.

The phase function $\phi_i(k)$ can be deduced directly from the phase of a complex
Fourier transform when the contribution of a given coordination shell (normally the
first shell) gives rise to an isolated peak in an r-space Fourier transform of the EXAFS
spectrum. This peak is transformed back to k-space after setting to zero all Fourier
components not part of the single shell peak. The phase or argument of the resulting
k-space backtransform is given by $2kr + \phi(k)$, where r is the distance to the isolated
shell. When r is known, as for a reference compound of known structure, $\phi(k)$ is
given by the backtransform phase minus $2kr$. This procedure is illustrated in Fig. 4.

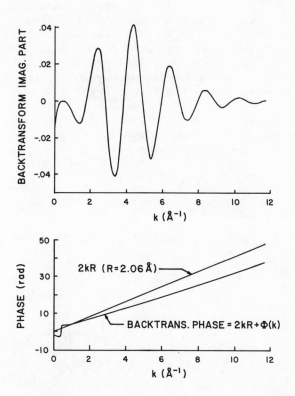

Fig. 4. Illustration of procedure for determination of the phase function $\phi(k)$ from the complex backtransform of a peak in the Fourier transformed EXAFS spectrum. Data are for a $Ni(NO_3)_2 \cdot 6H_2O$ solid reference compound.

When a single coordination shell cannot be isolated as described, $\phi(k)$ may be determined by fitting the model EXAFS function $\chi(k)^{21}$ to the data with r_i's fixed at values known for the reference compound, and using the envelope function of the reference spectrum itself to eliminate dependence on $F_i(k)$ and σ_i^2. Parameters in a representation of $\phi(k)$ are obtained from this fitting procedure.

In Fig. 5 results for these two procedures are compared for the case of an $Ni(NO_3)_2 \cdot 6H_2O$ hydrated $Ni(II)$ salt in which the Ni^{++} ion is octahedrally coordinated by water molecules.[30] Also shown in Fig. 5 is the theoretical calculation of $\phi(k)$ for the Ni-O combination by Teo and Lee.[31] Although there is close agreement between the three curves in Fig. 5, there appears to be a problem with the theoretical result in that the +2 valence state of the $Ni(II)$ ion has not been taken into account, and would be expected to move the theoretical curve positively, decreasing agreement with the empirical curves. The situation is not improved by choosing a different photoelectron energy reference value E_o for the theoretical calculation, because the main influence of this parameter is at lower k values. The effect on the theoretical results of the chemical state of the oxygen backscatterer (i.e. combined in water, or in a negative valence state) has not yet been investigated and may play a role in the observed discrepancy between theoretical and empirical phase functions.

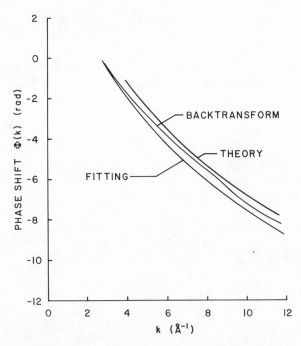

Fig. 5   Comparison of phase function $\phi(k)$ for $Ni-O$ in $Ni(NO_3)_2 \cdot 6H_2O$ as determined by the backtransform procedure, fitting of a model function to the Fourier filtered EXAFS spectrum and as predicted by theory (B. K. Teo and P. A. Lee, J. Am. Chem. Soc. **101**, 2815 (1979).

Because the envelope function for the EXAFS due to a given reference substance coordination shell consists of the backscattering amplitude $F_i(k)$ modulated by the Debye-Waller factor $exp(-2k^2\sigma_i^2)$, neither of these quantities can be determined separately without additional information. In practice it is convenient to use the product of these factors in fitting unknown spectra, and determine changes in σ_i^2 relative to the σ_i^2 value of the unknown. This is discussed in more detail elsewhere.[18]

Determination of coordination distances can be approached via the Fourier back-transform phase for cases in which this can be determined for single coordination shell (i.e. when peaks are isolated in the r-space Fourier transform). In this procedure, a phase shift function $\phi(k)$ is calculated for the unknown substance using an assumed r value and compared to $\phi(k)$ for the appropriate reference. In practice both r and E_o need to be varied to obtain a fit between the unknown and reference phase functions, as illustrated in Fig. 6.[32] An analogous procedure has been followed by Martens et al.[33]

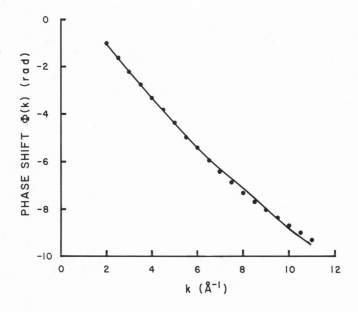

Fig. 6.   Illustration of phase fitting procedure for determination of bond lengths in a Ni^{++} 0.1 M solution. The phase function for solution (solid line) has been made to fit the phase function for the $Ni(NO_3)_2 \cdot 6H_2O$ reference solid (dots) by adjusting r and E_0, with results given in Table 1.

Dilute Aqueous Solutions

The phase fitting procedure described above has been applied to aqueous solutions of Ni^{2+}, Co^{2+}, and Cr^{3+} at 0.1 M concentration with the results given in Table 1,[32] along with x-ray scattering results for the same solutions. Bond lengths agree for the two techniques, but EXAFS results have been obtained at much lower concentrations. In the work described, direct absorption measurements were used, so that the concentration level of 0.1 M is by no means a lower limit.

Strong I-II Electrolytes

Concentration dependent local ordering is thought to occur in strong I-II electrolytes. For example, neutron diffraction studies of $NiCl_2-D_2O$ solutions by Howe et al.[5] identify an $Ni^{2+} - Ni^{2+}$ correlation, suggesting long range order. In addition, over a range of concentration, neutron diffraction has found Ni^{2+} to be approximately sixfold coordinated with an $Ni-O$ distance of 2.07 -2.10Å.[6,7] $Ni-D$ separation was also measured and found to vary with concentration, suggesting distortion of the hydration sphere at higher concentrations.

Raman spectroscopy[9,10] and transport measurements[11] all suggest formation of an ordered structure in $NiCl_2$ solutions above $\sim 1M$.

EXAFS measurements[16] for concentrated $CuBr_2$ solutions indicate a local coordination for Cu^{2+} similar to the hydrated salt, in which Cu^{2+} is primarily Br^- coordinated. EXAFS spectroscopy was applied to concentrated $NiCl_2$ solutions by Sandstrom.[18] EXAFS complements the neutron scattering technique in this application because it does not detect H or D, but is increasingly sensitive to the other scatterers present in the order O, Cl, Ni. In contrast, neutron scattering is most sensitive to D and least sensitive to Cl.[6]

In this case the r-space Fourier transform of the EXAFS spectra did not exhibit separate peaks, so that multiparameter fits of the model EXAFS function $\chi(k)$ were made to the experimental EXAFS spectra, using phase shift and amplitude functions obtained as described above.

Identification of a given scattering shell as oxygen or chlorine was facilitated by the observation that the phase shift functions for these two elements differ by approximately π over the full k range of interest. This makes these elements relatively distinguishable in the fitting process and causes attempts to fit with the wrong element to lead to negative coordination numbers. This point is illustrated in Fig. 7, which shows theoretically derived $Ni-O$ and $Ni-Cl$ phase functions vs. k.[31]

The results of this analysis are summarized in Table 2. The $NiCl_2$ solutions studied were found to have the Ni^{2+} ion sixfold water coordinated at 2.07 $\pm$ 0.01Å and coordinated by an outer sphere containing $\sim$3 Cl^- ions at 3.1 $\pm$0.03Å. No evidence was found for first sphere halogen coordination of the cation. The higher than stoichiometric average Cl^- coordination of Ni^{2+} may suggest bridging of Ni^{2+} coordination groups.

Recently Lagarde et al.[20] have studied I-II electrolyte solutions involving Sr, Cu, Ni, and Zn cations and Cl and Br anions. There appears to be general agreement with the $NiCl_2$ results described above.

Table 1

Results of Phase Fitting Analysis of EXAFS Spectra for Aqueous
Solutions in Comparison with X-ray Scattering

Technique	Ion	Concentration (mol/l)	1st Shell Radius (Å)	Reference
EXAFS	Ni^{2+}	0.1	2.06	32
X-ray Scatt.	Ni^{2+}	2.49	2.04	1
EXAFS	Co^{2+}	0.1	2.08	32
X-ray Scatt.	Co^{2+}	2.67	2.08	1
EXAFS	Cr^{3+}	0.1	2.01	32
X-ray Scatt.	Cr^{3+}	1.0	2.00	3

Fig. 7. Comparison of phase function $\phi(k)$ for $Ni-O$ and $Ni-Cl$ absorber scatterer combinations vs k, as predicted by theory (Ref. 31).

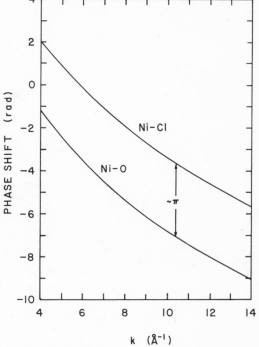

Table 2

Results of Least Squares Fits to $NiCl_2$ Solution EXAFS Spectra[*]

Concentration[a]	Shell	Element	N ($\pm 20\%$)	r (Å)	$\Delta \sigma^2$ (Å^2) $+-10\%$
		Two-Shell Fits			
2.78	1	O	7.0	2.06 ± 0.01	0.0018
	2	Cl	3.7	3.11 ± 0.03	0.013
3.74	1	O	6.2	2.07 ± 0.01	0.0011
	2	Cl	3.3	3.11 ± 0.03	0.013
		Three-Shell Fits			
2.78	1	O	7.0[b]	2.06 ± 0.01[b]	0.0018[b]
	2	O	0.5	2.90 ± 0.03	-0.0053
	3	Cl	2.8	3.11 ± 0.03	0.012
3.74	1	O	6.2[b]	2.07 ± 0.01[b]	0.0011[b]
	2	O	0.4	2.93 ± 0.03	-0.0073
	3	Cl	5.5	3.09 ± 0.03	0.025

[*]From Reference 18. [a]Molar concentration, gram-moles/liter. [b]Parameters not varied in three-shell fits.

Homogeneous Catalyst Studies

Cationic rhodium(I) complexes with optically active phosphine ligands are good catalysts for the hydrogenation of prochiral olefins to produce optically active products, such as L-DOPA.[34] Enantiomeric excesses as high as 99% have been reported.[35] In recent years a variety of techniques have been used to study the mechanism of these catalytic reactions, trying to identify and characterize the active catalysts in solution. Our most recent efforts have used single-crystal x-ray,[22] $^{13}C-$ and $^{31}P-NMR$ (solution), EXAFS,[22] and solid-state $^{13}C-NMR$ to characterize these catalysts of the general formulations: $Rh(P-P)L^+BF_4^-$ [36] or $Rh(P)_2L^+BF_4^-$.[36] A perspective drawing of one catalyst, $Rh(Dipamp)COD^+BF_4^-$ is seen in Fig. 8. Our work and the work of others,[37,38] studying the asymmetric hydrogenation of substrates such as (Ac) α-acyl- or α-benzoylaminocinnamic acids, has resulted in the proposed catalytic cycle shown in Fig. 9. Following is a summary of our EXAFS studies of these rhodium catalysts.

EXAFS studies of these rhodium catalysts can best be discussed in three areas: (1) comparisons of the structural parameters derived from EXAFS with the parameters determined by single crystal x-ray crystallography, (2) comparisons of the structural parameters derived from solid-state and solution EXAFS, and (3) studies of the active catalysts in solution. The results from the first two areas of study are summarized in Table 3. Before commenting on each area of study, a brief discussion of data analysis procedures will be given. In general, the procedures detailed by Sandstrom[18] were followed. One-shell to four-shell fits were used in the least-squares fitting procedures. Solid-state EXAFS data for Wilkinson's catalyst $Rh(P(C_6H_5)_3)_3Cl$, $Rh(Dipamp)COD^+BF_4^-$, and $Rh(cis-ethylene)COD^+BF_4^-$ were used to derive the various coefficients for Rh, P, C, and Cl. These coefficients were laboriously adjusted until one set of parameters for each atom type could be used to gain acceptable agreement for the 1st shell coordination about the rhodium. Oxygen contributions were obtained from analysis of the EXAFS data for Rh_2O_3. The derived parameters for Rh, P, C, and O were used in the least-squares fitting procedures to obtain the data given in Table 3. In the final analysis only N, r, and σ were varied. For systems requiring four-shell fits, only the parameters for two of the four shells were varied in any given cycle of refinement.

Comparison of Solid-State EXAFS With Crystallographic Data

The first portion of the EXAFS work was to compare the rhodium 1st shell coordination environments as determined by EXAFS and x-ray crystallography. As seen in the first two columns of Table 3, the EXAFS derived coordination distances agree very well with the crystallographic results. The accuracy of the $Rh-X$ bond distances as determined by EXAFS are in the range ±0.01-2Å. Single crystals of $Rh(Dipamp)Ac$ were not obtained, but based on the EXAFS results the structure for $Rh(diphos)Ac$ and $Rh(Dipamp)Ac$ are viewed as being identical in the solid-state.

Comparison of the Solid-State and Solution EXAFS

Our second interest was the comparison of structural parameters derived from EXAFS measurements for these rhodium catalysts as solids and solutions. The solution data were recorded after dissolving the crystalline compounds in methanol using the reaction cell described in the experimental section. In general, good agreement is obtained between the solid-state and solution data as seen in columns two and three of Table 3. The accuracy of the bond distances determined from the solution EXAFS data

Fig. 8. A perspective drawing showing the coordination environment about the rhodium atom in Rh (Dipamp) $COD^+BF_4^-$. The BF_4^- anion has been omitted for clarity.

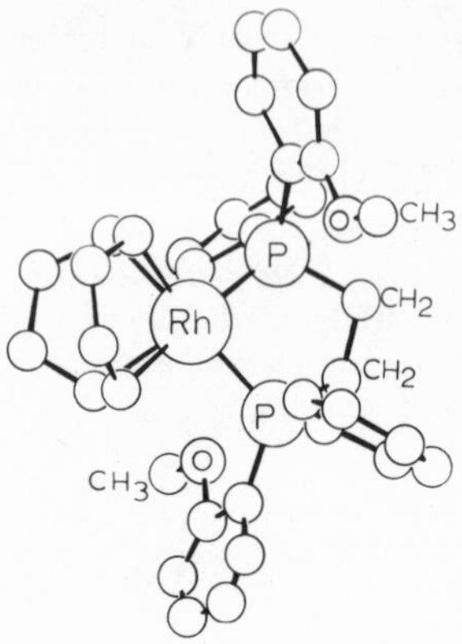

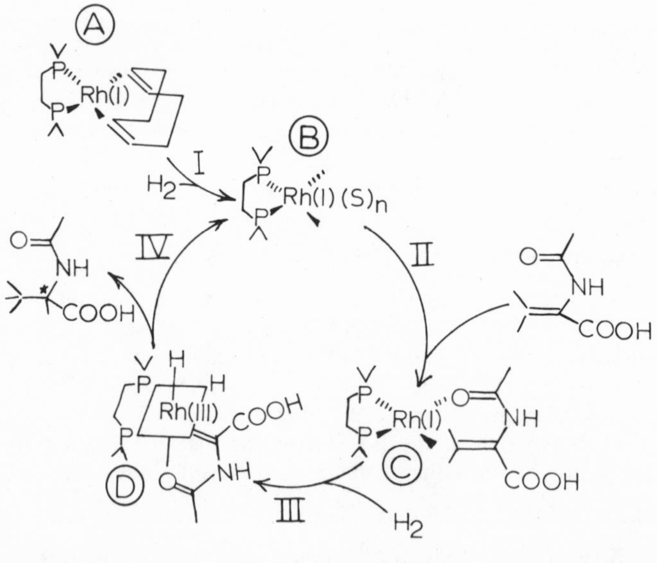

Fig. 9. The proposed catalytic cycle for the hydrogenation of prochiral olefins using asymmetric rhodium catalysts.

Table 3

Bond Distances from Least Squares Fits to Rhodium Catalysts EXAFS Spectra

Compound[a]		Single-Crystal X-Ray	Average Distances,Å EXAFS-Solid	EXAFS-Soln
Rh(Diphos)COD[+e]	Rh-P[b]	2.28	2.29	2.33
	Rh-C	2.25	2.22	2.32
Rh(Dipamp)COD[+e]	Rh-P[b]	2.27	2.28	2.28
	Rh-C	2.20	2.22	2.34
Rh(Diphos)Ac[h,i]	Rh-P[c]	2.27	2.27	---
	Rh-P	2.23	2.23	---
	Rh-C	2.22	2.28	---
	Rh-O	2.11	2.12	---
Rh(Dipamp)Ac[i]	Rh-P[c]	---	2.29	---
	Rh-P	---	2.26	---
	Rh-C	---	2.28	---
	Rh-O	---	2.01	---
Rh(Dipamp)[g]$_2$	Rh-P[d]	2.31 (Diphos)	2.30	2.29
Rh(AcAc)[f]$_3$	Rh-O[d]	1.99	2.03	2.07
Rh(Diop)NBD[+e]	Rh-P[b]	2.30	2.29	2.34
	Rh-C	2.22	2.19	2.36
Rh(Camp)$_2$COD[+e]	Rh-P[b]	2.35	2.32	2.34
	Rh-C	2.20	2.10	2.30
Rh(Cis-ethylene)COD[+e]	Rh-P[b]	2.26	2.27	2.34
	Rh-C	2.26	2.26	2.30

[a]See reference 36 for ligand abbreviations. [b]Two shell fit. [c]Four shell fit. [d]One shell fit. [e]X-ray crystallographic results, see reference 22. [f]X-ray crystallographic results, see: J. C. Morrow and E. B. Parker, Acta. Cryst., B29, 1145 (1973). [g]X-ray crystallographic results, see: M. C. Hall, B. T. Kilbourn, and K. A. Taylor, J. Chem. Soc. (A), 2540 (1970). [h]See reference 37. [i]Ac is α-acylaminocinnamic acid.

is of the order ± 0.03-5Å. This is due in part to the loss of data in the high k region. In Fig. 10 the raw EXAFS data after background corrections for $Rh(Dipamp)COD^+$ in solution (solid line) is compared to the solid-state data (dashed line). Very little data above $k = 14\text{Å}^{-1}$ is seen for the solution, whereas, the data in the solid-state extend well above $k = 18\text{Å}^{-1}$. A general increase in $Rh-P$ and $Rh-C$ bond distances is seen in going from solid to solution. This could be due to relaxation of the crystal packing forces, a failure of our fitting parameters (as derived from solid-state data) to adequately describe the solution EXAFS data, or a combination of both effects.

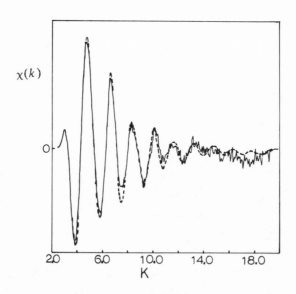

Fig. 10.   EXAFS data $\chi(k)$ vs k in Å^{-1} for $Rh(Dipamp)COD^+BF_4^-$ as a crystalline solid (dashed line) and in solution (solid line). The ordinate scale is -8 to 8.

Studies of Active Catalyst in Solution

The third part of our study was to determine structural parameters for the catalyst under simulated reaction conditions. Also of interest was the identification of the intermediates in the catalytic cycle shown in Fig. 9. Our work in this area is by no means complete but the results look very promising. We have determined the structural parameters for species B (the active catalysts in solution, Fig. 9) for two different phosphine ligand systems. The results are given in Fig. 11. The first shell coordination environment about the rhodium atoms consist of two phosphorus atoms and two methanol oxygen atoms. Attempts to force fit the data having different numbers of oxygen atoms bonded to the rhodium are shown in Fig. 12. We also attempted to fit the data with two phosphorus and four carbon atoms bonded to the rhodium. In all cases the fit obtained with two phosphorus and two oxygen atoms was superior. Data have been obtained and analyzed for species C in Fig. 9 where the phosphorus ligand was Dipamp. When the substrate (α-acylaminocinnamic acid, Ac) was added to the methanol solvated species, $Rh(Dipamp)(CH_3OH)_2^+$, the following structural parameters were obtained: $Rh-P_1 = 2.36\text{Å}$, $Rh-P_2 = 2.29\text{Å}$, $Rh-C = 2.26\text{Å}$, and $Rh-O = 1.86\text{Å}$. There are two distinct $Rh-P$ distances, one being trans to an oxygen and one trans to the olefin. This square planar arrangement is supported by the x-ray structural analysis of Halpern.[37] The coordination environment about the rhodium is $\begin{smallmatrix} P \diagdown & \diagup C \\ & Rh-C \\ P \diagup & \diagdown O \end{smallmatrix}$. The $Rh-C$ and $Rh-P$ distances derived from the EXAFS data for the in-situ prepared catalyst-substrate complex agree favorably with the values for the crystalline $Rh(Dipamp)Ac^+$ compound in Table 3. The short $Rh-O$ distance of 1.86Å compared to $>2.0\text{Å}$ for the crystalline compounds is surprising and may not be valid. However, recent experimental observations[39] indicate differences between the crystalline $Rh(Dipamp)Ac^+$ and the in-situ generated species in solution. More EXAFS data for other Rh-phosphine-substrate complexes is needed to verify the short $Rh-O$ distance.

The sensitivity of EXAFS data to different species in solution is illustrated in the following experiment. When the catalyst substrate was added to the di-solvated species $Rh(Camp)_2(CH_3OH)_2^+$ ($Camp$ is a monodentate ligand) no reasonable information pertaining to the catalyst-substrate complex was obtained. Attempts to fit the data as either the di-solvated species or some reasonable catalyst-substrate complex resulted in diverging least-squares refinements. Numerous attempts to refine these data were unsuccessful. In hindsight, this was to be expected. From NMR experiments we now know that when the phosphine ligands are monodentate, as for Camp, rapid isomerization occurs in solution. This isomerization is not possible for bidentate ligands such as Dipamp. At best we would expect a mixture of three or four different species in solution after the substrate was added. The fact that we were unable to force fit this data is important in illustrating the sensitivity of EXAFS data to different species in solution.

In summary, our work with the asymmetric rhodium catalysts, as well as others with polymer-bound Wilkinson's catalysts[40] has shown EXAFS to be a valuable tool for studying homogeneous catalysis. We obtained good agreement between the structural parameters determined by EXAFS and x-ray crystallography. Our work demonstrates that useful information pertaining to the identity and structural parameters for solutions may be obtained from EXAFS. A word of caution is that in trying to distinguish subtle differences between species using EXAFS, one needs to have very good model compounds whose structures are known.

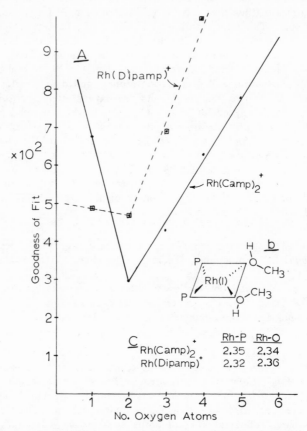

Fig. 11.   (a) Plots of the goodness of fit from a two shell least squares fit vs the number of oxygen atoms in the fit. For all fits two phosphorus atom contributions were included. (b) The proposed structure in solution. (c) Coordination distances derived from the best two shell fits of the solution EXAFS data (see Reference 36 for ligand abbreviations).

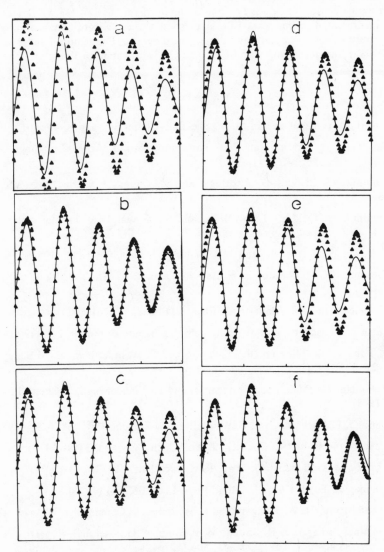

Fig. 12   Solid curves: The Fourier filtered and k^3 weighted EXAFS spectrum for Rh (CAMP)$_2^+$ in methanol. Triangles: Two shell fit to the solid curve. All fits include contributions from two phosphorus atoms plus in (a) one oxygen atom, (b) two oxygen atoms, (c) three oxygen atoms, (d) four oxygen atoms, (e) five oxygen atoms, and (f) four carbon atoms. The scales are $k = 4$ to 12Å^{-1} for the abscissas and $\chi(k)k^3 = -0.6$ to 0.6 for the ordinates.

ACKNOWLEDGEMENTS

The authors wish to thank F. W. Lytle for helpful discussions. Part of this work was supported by the National Science Foundation under grant CHE77-22315. Some of the materials incorporated in this work were developed at the Stanford Synchrotron Radiation Laboratory which is supported by the National Science Foundation (under contract DMR77-27489), in cooperation with SLAC and the Department of Energy.

REFERENCES

1. H. Ohtaki, T. Yamaguchi, and M. Maeda, Bull. Chem. Soc. Japan. **49**, 701 (1976).

2. W. Bol, G. J. A. Gerrits, and C. L. van Panthaleon, J. Appl. Crystallogr. **3**, 486 (1970).

3. R. Caminiti, G. Licheri, G. Piccaluga, and G. Pinna, J. Chem. Phys. **65**, 3134 (1976).

4. R. Caminiti, G. Licheri, G. Piccaluga, and G. Pinna, Disc. Faraday Soc. **64**, 62 (1977).

5. R. H. Howe, W. S. Howells, and J. E. Enderby, J. Phys. **C7**, L111 (1974).

6. A. K. Soper, G. W. Neilson, J. E. Enderby, and R. A. Howe, J. Phys. **C10**, 1793 (1977).

7. G. W. Neilson and J. E. Enderby, J. Phys. **C11**, L625 (1978).

8. G. Cubiotti, F. Sachetti, and M. C. Spinelli, Solid State Commun. **27**, 349 (1978).

9. M. P. Fontana, G. Maisano, P. Migliardo, and F. Wanderlingh, Solid State Commun. **23**, 489 (1977).

10. M. P. Fontana, G. Maisano, P. Migliardo, and F. Wanderlingh, J. Chem. Phys. **69**, 676 (1978).

11. G. Maisano, P. Migliardo, and F. Wanderlingh, J. Chem. Phys. **68**, 5594 (1978).

12. J. W. Neely and Robert E. Connick, J. Am. Chem. Soc. **94**, 3419 (1972).

13. H. Winick and A. Bienenstock, Ann. Rev. Nucl. Part. Sci. **28**, 33 (1978).

14. P. Eisenberger, B. M. Kincaid, Chem. Phys. Lett. **36**, 134 (1975).

15. D. R. Sandstrom, H. W. Dodgen, and F. W. Lytle, J. Chem. Phys. **67**, 473 (1977).

16. A. Fontaine, P. Lagarde, D. Raoux, M. P. Fontana, G. Maisano, P. Migliardo, and F. Wanderlingh, Phys. Rev. Lett. **41**, 504 (1978).

17. T. I. Morrison, A. H. Reis, G. S. Knapp, F. Y. Fradin, H. Chen, and T. E. Klippert, J. Am. Chem. Soc. **100**, 3262 (1978).

18. D. R. Sandstrom, J. Chem. Phys. **71**, 2381 (1979).

19. H. W. Huang, S. H. Hunter, W. K. Warburton, and S. C. Moss, Science **204**, 191 (1979).

20. P. Lagarde, A. Fontaine, D. Raoux, A. Sadoc, and P. Migliardo, Private Communication.

21. D. R. Sandstrom and F. W. Lytle, Ann. Rev. Phys. Chem. **30**, 215 (1979).

22. B. R. Stults, Presented at Amer. Chem. Soc. — Chem. Soc. Japan Chemical Congress, Honolulu, Hawaii, April 1979.

23. G. S. Knapp, H. Chen, and T. E. Klippert, Rev. Sci. Instrum. **49**, 1658 (1978).

24. S. Doniach, I. Lindau, W. E. Spicer, and H. Winick, J. Vac. Sci. Tech. **12**, 1123 (1975).

25. J. Jaklevic, J. A. Kirby, M. P. Klein, A. S. Robertson, G. S. Brown, and P. Eisenberger, Solid State Comm. **23**, 679 (1977).

26. S. C. Moss, H. Metzger, M. Eisner, H. W. Huang, and S. C. Hunter, Rev. Sci. Instru. **49**, 1559 (1978).

27. E. A. Stern and S. H. Heald, Rev. Sci. Instru. **50**, 1579 (1979).

28. P. Eisenberger and B. M. Kincaid, Science **200**, 1441 (1978).

29. E. A. Stern, Contemp. Phys. **19**, 289 (1978).

30. F. Bigoli, A. Braibanti, A. Tiripicchio, and M. Tiripicchio-Camellini, Acta Crystallogr. Sect. B **27**, 1427 (1971).

31. B. K. Teo and P. A. Lee, J. Am. Chem. Soc. **101**, 2815 (1979).

32. D. R. Sandstrom and J. M. Fine, SSRL Rep. 78/09, Stanford Synchrotron Radiation Lab., Palo Alto, Calif. (1978), pp. 77-78.

33. G. Martens, P. Rabe, N. Schwenterner, and A. Werner, Phys. Rev. B **17**, 1481 (1978).

34. For example see, B. D. Vineyard, W. S. Knowles, M. J. Sabacky, G. K. Bachman, and D. J. Weinkauff, J. Amer. Chem. Soc. **99**, 5946 (1977).

35. M. D. Fryzuk and B. Bosnich, J. Amer. Chem. Soc. **99**, 6262 (1977).

36. Ligand abbreviations: P-P is Dipamp = (R,R)-1,2-ethanediylbis-[(O-methoxyphenyl)phenylphosphine], Diop = (R,R)-isopropylidene-2,3-dihydroxy-1,4-bis-(diphenylphosphino)butane, Diphos = bis(1,2-diphenylphosphino)ethane, Cis-ethylene = cis-bis-(Diphenylphosphino)ethylene; P is Camp = (R)-O-anisylmethyl-cyclohexylphosphine; L = 1,5-cyclooctadiene or bicyclo-[2.2.1]-heptadiene.

37. J. Halpern, Trans. Amer. Cryst. Assoc., March 20, 1978, pp. 59.

38. J. M. Brown and P. A. Chaloner, Tetrahedron Lett., 1988 (1978) and J. M. Brown, private communication.

39. J. Halpern, University of Chicago, private communication.

40. J. Reed, P. Eisenberger, B. K. Teo, and B. M. Kincaid, J. Am. Chem. Soc. **99**, 5217 (1977); *ibid.* **100,** 2375 (1978).

EXAFS STUDIES OF SUPPORTED METAL CATALYSTS

G. H. Via and J. H. Sinfelt

Exxon Research and Engineering Co.
Linden, New Jersey 07036

and

F. W. Lytle

The Boeing Company
Seattle, Washington 98124

EXAFS (Extended X-Ray Absorption Fine Structure) is an element specific electron scattering technique in which a core electron ejected by an x-ray photon probes the local environment of the absorbing atom. The ejected photoelectron is backscattered by the neighboring atoms around the absorbing atom and interferes constructively or destructively with the outgoing electron wave, depending on the energy of the photoelectron. The energy of the photoelectron is equal to the difference between the x-ray photon energy and a threshold energy associated with ejection of the electron. In the EXAFS experiment the photoelectron energy is varied by varying the energy of the incident x-ray beam. The interference between outgoing and backscattered electron waves has the effect of modulating the x-ray absorption coefficient so that absorption spectra determined experimentally exhibit oscillations in the absorption coefficient on the high energy sides of absorption edges. The electron wave vector, k, is related to the kinetic energy, E, and wavelength, λ_e, by

$$k = (2mE)^{1/2}/\hbar = 2\pi/\lambda_e \qquad (1)$$

where m is the mass of the electron and $\hbar$ is Planck's constant divided by 2π. The EXAFS function, $\chi(k)$, is defined in terms of atomic absorption coefficients,

$$\chi(k) = (\mu - \mu_o)/\mu_o \qquad (2)$$

where μ refers to absorption by an atom in a material of interest and μ_o refers to absorption by the atom in the free state.

159

Theories of EXAFS[1] based on scattering of the ejected photoelectron by atoms in the immediate vicinity of the absorbing atom give an expression for $\chi(k)$ of the form

$$\chi(k) = \sum_j A_j(k) \, sin\,[2kr_j + 2\delta_j(k)] \qquad (3)$$

where the summation extends over j coordination shells. r_j is the radial distance from the absorbing atom to atoms in the jth shell and $2\delta_j(k)$ is the phase shift. The amplitude function, $A_j(k)$, is given by

$$A_j(k) = (N_j/kr_j^2) \, F(k) \, exp\,(-2r_j/\lambda) \, exp\,(-2k^2\sigma_j^2) \qquad (4)$$

In this expression N_j is the number of atoms in the jth shell, σ_j is the rms deviation of distance about r_j, $F(k)$ is the backscattering amplitude and λ is the mean free path for inelastic scattering. The backscattering factor $F(k)$ depends on the kind of atom responsible for the scattering.[2]

The analysis of EXAFS data for structural information (N, σ, r) generally proceeds by the use of Fourier transform and non-linear least squares techniques. In the case of highly dispersed metal catalysts, we have been most interested in information on metal atoms in the first coordination shell. For monometallic clusters this shell will contain only one type of backscattering atom while for bimetallic clusters two types of atoms will be present. To limit the problem to the first shell of metal atoms, the EXAFS function, $\chi(k)$, is Fourier transformed to produce a radial structure function, $\phi(r)$.[3] A small region of $\phi(r)$ corresponding to the r range containing the first shell of metal atoms is then isolated and transformed back into k space, and the resulting filtered EXAFS function is analyzed to obtain structural parameters of interest using non-linear least squares techniques. To further limit the number of unknowns in the problem, information obtained on reference materials is used where appropriate. For example, in the case of monometallic catalysts, data obtained on pure metal foils at the same temperature as the catalyst can be used to extract the phase shift function $2\delta_1(k)$ and the product $F(k) \cdot exp\,(-2r_1/\lambda) \cdot exp\,(-2k^2\sigma_1^2)$. With this information, the only unknowns in the problem are N_1, r_1, and $\Delta\sigma_1^2$, where $\Delta\sigma_1^2$ is the difference in σ_1^2 between the standard and the unknown at the temperature of measurement, which may be either positive or negative depending on whether the value of σ_1^2 in the unknown is larger or smaller than in the reference material.

We have utilized EXAFS to study the average structure of the metal clusters in a number of highly dispersed monometallic catalysts.[4] Clusters of osmium, iridium and platinum dispersed on silica or alumina were investigated. The metal clusters in the catalysts constituted 1.0 wt.% of the total mass, and measurements were made following thorough reduction of the catalysts at elevated temperature. The EXAFS data on the metal clusters were compared to data on the corresponding bulk metals in the analysis. The general conclusions of this work are as follows:

1. Nearest neighbor distances in the metal clusters differed from those in the bulk metals by less than 0.02Å.

2. Nearest neighbor coordination numbers for the metal clusters ranged between 7 and 10 as compared to 12 for the bulk metals. These low coordination numbers are consistent with the high state of dispersion of the catalysts.

3. The value of the parameter σ_1 for pairs of atoms in the metal clusters was found to be larger by a factor of 1.4 to 2 than σ_1 for the bulk metals, which reflects the high percentage of surface atoms in the metal clusters.

Most recently, we have applied these data analysis techniques in the determination of the average structure of supported bimetallic cluster catalysts such as silica supported ruthenium-copper (1 wt.% *Ru* - 0.63 wt.% *Cu*). Catalytic data on ethane hydrogenolysis activity and on selectivity of cyclohexane conversion to benzene,[5] plus H_2 chemisorption data in conjunction with electron microscopy results,[6] and preliminary EXAFS data[7] have demonstrated that copper and ruthenium strongly interact despite the fact that these two metals show very limited bulk miscibility. These results indicate that the bimetallic clusters contain a copper-rich surface, and a detailed analysis of the EXAFS data supports this interpretation.

EXAFS data on both the copper and ruthenium components of the system were utilized so that the local environment of each type of atom could be separately investigated. The data were obtained at 100°K on catalysts which had been reduced *in situ* with flowing hydrogen at 700°K using a previously described[8] furnace. In the analysis a detailed comparison was made between pure silica supported ruthenium (1 wt.% metal) and the ruthenium component of the bimetallic clusters, and between pure silica supported copper (0.63 wt.% metal) and the copper component of the bimetallic clusters.[9] The results of this comparison are that ruthenium has about 90% ruthenium neighbors and 10% copper neighbors in its first coordination shell, while copper has about 50% copper neighbors and 50% ruthenium neighbors in its first coordination shell. These results are consistent with a model of the bimetallic clusters in which a ruthenium rich core is bounded by a thin layer of chemisorbed copper. Evidence was also found that $Ru-Ru$ and Cu-Cu interatomic distances in the bimetallic clusters were slightly different from the values in bulk ruthenium and copper.

Additional evidence for a copper-rich surface in the supported bimetallic clusters comes from EXAFS studies of the bimetallic clusters and pure metal clusters in the presence of chemisorbed oxygen. These studies show that copper in the bimetallic clusters interacts much more extensively with oxygen than copper in the pure silica supported copper catalyst. At the same time, ruthenium in the bimetallic clusters shows considerably less interaction with oxygen than ruthenium in the pure silica supported ruthenium catalyst, even though the average cluster size of the bimetallic clusters is slightly lower than the average cluster size of the pure ruthenium clusters.[6] These data suggest that copper concentrates in the surface of the bimetallic clusters and thus significantly isolates the ruthenium from oxygen interaction.

ACKNOWLEDGEMENT

We are grateful to the excellent staff at Stanford Synchrotron Radiation Laboratory for experimental support and to DOE and NSF for support of the facility. The research of FWL was partially supported by NSF Grants CHE 76-11255 and DMR 77-12919.

REFERENCES

1. (a) D. E. Sayers, F. W. Lytle, and E. A. Stern, Phys. Rev. Lett. **27**, 1204 (1971); (b) D. E. Sayers, F. W. Lytle, and E. A. Stern, in *Advances in X-Ray Analysis,* ed. by B. L. Henke, J. B. Newkirk and G. R. Mallett, Plenum, New York, 1970, Vol. 13, p. 248; (c) E. A. Stern, Phys. Rev. B **10**, 3027 (1974); (d) C. A. Ashley and S. Doniach, Phys. Rev. B **11**, 1279 (1975); (e) P. A. Lee and J. B. Pendry, Phys. Rev. B **11**, 2795 (1975); (f) P. A. Lee and G. Beni, Phys. Rev. B **15**, 2862 (1977).

2. (a) B. K. Teo, P. A. Lee, A. L. Simons, P. Eisenberger, and B. M. Kincaid, J. Am. Chem. Soc., **99**, 3854 (1977); (b) B. K. Teo and P. A. Lee, J. Am. Chem. Soc., **101**, 2815 (1979).

3. D. E. Sayers, Ph.D. Dissertation, University of Washington, 1971.

4. (a) J. H. Sinfelt, G. H. Via, and F. W. Lytle, J. Chem. Phys. **68**, 2009 (1978); (b) G. H. Via, J. H. Sinfelt, and F. W. Lytle, J. Chem. Phys. **71**, 690 (1979).

5. (a) J. H. Sinfelt, J. Catal. **29**, 308 (1973); (b) J. H. Sinfelt, Rev. Mod. Phys. **51**, 569 (1979).

6. E. B. Prestridge, G. H. Via, and J. H. Sinfelt, J. Catal. **50**, 15 (1977).

7. F. W. Lytle, G. H. Via, and J. H. Sinfelt, PREPRINTS, Div. of Petrol. Chem., ACS, **21**(2), 366 (1976).

8. F. W. Lytle, P. S. P. Wei, R. B. Greegor, G. H. Via, and J. H. Sinfelt, J. Chem. Phys. **70,** 4849 (1979).

9. J. H. Sinfelt, G. H. Via, and F. W. Lytle, J. Chem. Phys. **72,** 4832 (1980).

EXAFS OF AMORPHOUS MATERIALS

S. H. Hunter

Department of Applied Physics
Stanford University
Stanford, California 94305

The structure of amorphous alloys is an important field in materials research. The lack of an extensive regular structure in these amorphous materials, however, often limits the amount of unambiguous structural information which can be gained from conventional structural techniques such as x-ray, neutron, or electron diffractions. This problem is particularly severe for those systems which have small percentages of impurity atoms. Yet many amorphous alloy systems of interest are those with dilute impurities. In this chapter, we shall see that one can gain important structural information for these materials through careful application of a new structural tool called Extended X-Ray Absorption Fine Structure (EXAFS).

EXAFS is a technique which can probe the structure of atoms surrounding each species of atom in the material simply by tuning the x-ray energy to the absorption edge of that element.[1] With careful data acquisition and analysis EXAFS has the capability of providing interatomic distance, coordination number and degree of disorder not only for the majority atomic species in the material, but also for the very dilute components. For multi-component systems and for systems with dilute impurities, this specific information has not been obtained by other techniques. Since the primary restriction on the EXAFS sample is to have a uniform sample thickness of one to two absorption lengths, EXAFS can be useful in probing the structure of even single component amorphous systems under many different physical conditions. In addition, the high sensitivity of EXAFS to disorder could be used to compare degree of disorder in amorphous systems.

In amorphous materials, as in crystalline materials, the understanding of physical properties is often closely related to the structure or arrangement of atoms. In relating physical properties, such as electrical or thermal properties, to the structure, the nearest neighbor coordination number N and the coordination distance r are important as well as the atomic species Z of the neighboring atoms and the disorder, which can be either thermal, structural or compositional.

163

Conventional structural techniques provide limited information about amorphous materials. Radial distribution function (RDF) from x-ray or neutron scatterings is a measure of the average structural arrangement around an average atom in the material and yields ambiguous results when there are several different atom pairs in the material with similar coordination distances.[2] Anomolous scattering technique could resolve some of this ambiguity, but it is also limited since conventional x-ray sources have limited number of wavelengths and thus are only appropriate for certain kinds of materials. Anomolous scattering using synchrotron radiations, still under development, could widen the applicability of this technique in the future.[3]

The advantages of the EXAFS technique are that it can probe the near neighbor environment of each component species separately, that it can be used even with very dilute (milli-molar) impurity concentrations, and that it is more sensitive to disorder than are RDF's. The limitations of the EXAFS technique are: (1) the requirements of uniformly thin samples; (2) the need for empirical standards with similar known structures; and (3) the limitation of the study to only first nearest neighbors in amorphous materials due to the disorder.

EXAFS studies of amorphous systems, such as *Se* and *Ge* with known amorphous coordinations, have been used to test the accuracy of this technique and, for other amorphous systems, to solve structural questions.[4,5] The loss of the low k-range EXAFS data due to overlap with other physical phenomena occurring at the absorption edge, effectively limits the structural information obtained to just the first neighbor shell since this low-k region would Fourier transfer to the neighboring shells at higher distance r. This fact can be used to advantage in coordination number determination for amorphous materials since it naturally isolates the first neighbor shell in the Fourier transform from overlap with the second neighbor shell. However, an accurate determination still depends on good data acquisition and careful data analysis.

The determination of coordination distance has been reported many times for both crystalline and amorphous materials and will not be treated here. Rather, the accurate determination of coordination number, N, will be discussed and an example will be given for a multicomponent system with dilute impurities ($Cu-As_2Se_3$).[6] An amorphous system of interest and with no dilute impurities (*GeSe*) will also be considered briefly.

Obtaining good data depends first on having a steady, tunable x-ray beam with no harmonic content or with a detection system which minimizes the effects of harmonics, such as synchrotron radiation source with appropriate choice of detectors. In addition, the sample thickness should be homogeneous of one to two x-ray absorption lengths for good signal-to-noise, and, if possible, a measurable signal should be obtained over a long k-range, preferably at low temperature to minimize thermal disorder.

Careful data analysis requires normalization of the data, suitable choice of standard materials for comparison, and use of the same Fourier transform windows for both samples and standards being compared. Although not always necessary for coordination distance determination, normalization of the EXAFS oscillations is essential for accurate determination of coordination number and disorder calculations because they are calculated from the amplitude of the oscillations rather than the frequency, as we will see in the EXAFS equation. Normalization first requires removal of all non-K-shell absorption effects by fitting the absorption below the edge and subtracting the extrapolation of that from the absorption measured above the edge. The second step is normalization of the oscillations as a function of wave vector k by dividing the total K-shell absorption, $\mu(k)$, by the smooth (free-atom-like) absorption, $\mu_o(k)$, as given in the definition of the EXAFS oscillations:

$$\chi(k) \equiv \frac{\Delta\mu(k)}{\mu_o(k)} = \frac{\mu(k) - \mu_o(k)}{\mu_o(k)} \tag{1}$$

The structural information can then be obtained through fitting or through $(2kr)$ Fourier transform analysis of the EXAFS equation:[1]

$$\chi(k) = -\frac{1}{k}\sum_i \frac{N_i}{r_i^2} \sin(2kr_i+\phi_i)\, |F_i(k)|\, e^{-2\sigma_i^2 k^2}\, e^{-2r_i/\lambda} \tag{2}$$

where k=wave vector of the photoelectron ejected in the K-shell absorption, F=backscattering amplitude of neighboring atoms, ϕ=phase shift of photoelectron, σ^2 is a Gaussian measure of the disorder, $e^{-2r/\lambda}$ is a mean-free-path damping term for the photoelectron and the summation is over all shells of atoms.

The determination of coordination distances and numbers can be done either by fitting with calculated values for phase shifts and backscattering amplitudes[7] or by using Fourier transform techniques and comparison with empirical standard materials with a known structure similar to the sample.[1] The choice of the standard material is often simplified for amorphous systems by choosing the crystalline form of the material, since the chemical nature of the two forms is often very similar.

Also important for accurate determinations are the windows for the Fourier transform and the inverse Fourier transform. These must be chosen to minimize the truncation effects and to isolate the first neighbor peak and must be the same for both the sample and the standard.

The magnitude of the first neighbor peak in the Fourier transform should not be used to determine the coordination number N because N and σ^2 are strongly coupled and even a small disorder greatly reduces the peak height. Instead, N and σ^2 can be obtained by performing an inverse Fourier transform over just the first neighbor peak of both the sample and the standard (using a window), thereby producing the EXAFS oscillations due to just that first shell. A ratio of the magnitude of that inverse Fourier transform, $|\Lambda|$,

$$\frac{|\Lambda_\alpha|}{|\Lambda_\beta|} = \frac{N_\alpha}{N_\beta}\, \frac{r_\beta^2}{r_\alpha^2}\, \frac{F_\alpha(k)}{F_\beta(k)}\, \frac{e^{-2\sigma_\alpha^2 k^2}}{e^{-2\sigma_\beta^2 k^2}}\, \frac{e^{-2r_\alpha/\lambda_\alpha}}{e^{-2r_\beta/\lambda_\beta}} \tag{3}$$

Assuming that the sample and the standard have similar environments with the same backscattering amplitude, phase shifts, and mean free path, the log of this ratio becomes[1]

$$\ln\left(\frac{|\Lambda_\alpha|}{|\Lambda_\beta|}\right) = \ln\left(\frac{N_\alpha r_\beta^2}{N_\beta r_\alpha^2}\right) - 2\left(\sigma_\alpha^2 - \sigma_\beta^2\right) k^2 \tag{4}$$

This analysis assumes also that the difference in the disorder can be expressed as Gaussian. A plot of the log of the ratio versus k^2 would yield a straight line with the coordination number being obtained from the intercept (extrapolated to k=0) and the disorder (σ^2) from the slope.

The curvature that appears in some plots of this data has several possible sources. The Fourier transform truncation errors can be minimized by choice of the transform window. Detector system response to both beam harmonics and either sample pinholes or sample thickness inhomogeneities were predicted[8] to cause an erroneously low evaluation of N and this was later demonstrated experimentally.[9] Care was taken in sample preparation to reduce this. Another possible source of error is the assumption of only one kind of neighbor. Calculations of the effect of backscattering from different kinds of neighboring atoms produced s-shaped curvature in the ratio plot.[10]

This technique of coordination number determination was tested empirically by applying to systems of known coordination for both amorphous and crystalline forms: Se, Ge, and As_2Se_3. The results for these systems did have an accuracy of about half an atom out of three to four neighboring atoms.[5,6]

Cu-As$_2$Se$_3$ Systems

An example of coordination number determination in a multicomponent system with dilute impurities and suited to EXAFS analysis is the $a-Cu-As_2Se_3$ system.[2,6] In this system the addition of just 5 at% Cu to $a-As_2Se_3$ increases the electrical conductivity by four orders of magnitude, which is unusual in amorphous semiconductors. The previous structural studies (x-ray RDF, ESCA), on 0 to 30 at% Cu, determined the average coordination for an average atom but were limited from any more specific determination by the small percentage of the Cu and by very similar coordination distances.[2] This average coordination number was found to equal 8 minus the average number of valence electrons, as would occur in covalent bonding. EXAFS measurements were made at SSRL of the K-shell EXAFS for the Cu, As, and Se edges of amorphous alloys of 5 at% and 25 at% Cu in As_2Se_3, $c-As_2Se_3$ and $c-CuAsSe_2$. The $c-As_2Se_3$ has As atoms 3-fold coordinated with Se atoms and Se atoms 2-fold coordinated with As atoms, whereas the $c-CuAsSe_2$ has each Cu atom 4-fold coordinated with Se atoms, each As atom 3-fold coordinated with Se atoms and the Se atoms half 3-fold and half 4-fold coordinated. These materials were the two standards used in this study.

The analysis of the EXAFS data showed that the coordination distances in the amorphous samples were the same as in the crystalline samples, to within 0.02Å. No apparent second neighbor peaks were observed in the Fourier transform of the amorphous samples; thus the first neighbor peaks could be readily isolated from an inverse Fourier transform. A comparison of the EXAFS results for the amorphous and crystalline samples as a function of k^2 for the Cu, As and Se edges is given in Figures 1-3. Figure 1 shows that the Cu coordination number in the amorphous samples is approximately four as in the crystalline sample (intercept ≈ 0) with some increase in disorder as the Cu concentration increases.

The As edge comparison in Figure 2 shows that the 5 at % Cu sample (curve (a)) has 3-fold coordination (as in $c-As_2Se_3$) to within an accuracy of half an atom. The additional curvature for the 25 at % Cu sample can be evaluated as a coordination number between 3.5 and 5. This indicates an increase in the As coordination towards four-fold with the increasing Cu content. A later analysis of an amorphous structural model and of typical chemical bonding in $Cu-As-Se$ materials suggests that the addition of Cu not only increases the coordination number of the As atoms, but also introduces Cu or As atoms into the originally only-Se-atom first neighbor shell.[6] This

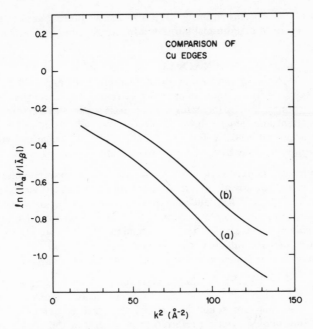

Fig. 1.   Comparison of nearest neighbor shells: $\ln(|\Lambda_\alpha|/|\Lambda_\beta|)$ versus k^2 for
(a) $\alpha = a - Cu_{.05}(As_{.4}Se_{.6})_{.95}$ $\beta = c - CuAsSe_2$
(b) $\alpha = a - Cu_{.25}(As_{.4}Se_{.6})_{.75}$ $\beta = c - CuAsSe_2$

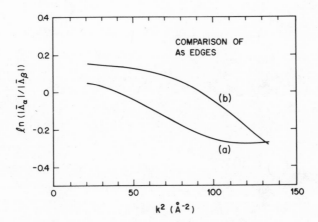

Fig. 2.   Comparison of nearest neighbor shells: $\ln(|\Lambda_\alpha|/|\Lambda_\beta|)$ versus k^2 for
(a) $\alpha = a - Cu_{.05}(As_{.4}Se_{.6})_{.95}$ $\beta = c - As_2Se_3$
(b) $\alpha = a - Cu_{.25}(As_{.4}Se_{.6})_{.75}$ $\beta = c - As_2Se_3$

change in kind of neighboring atom would introduce a different backscattering ampli-tude and could explain the additional curvature in the As-edge comparison of the 25 at % Cu sample.

The comparison of the Se EXAFS in Figure 3 for the two amorphous samples clearly shows an increase in the Se coordination number from 2-fold as in $c-As_2Se_3$ (curve (a), intercept ≈ 0) to approximately 4-fold (curve (b), intercept ≈ 0.9) coordi-nation as the Cu concentration increases from 5 to 25 at %. A structural model which maintains bond-count and which allows all the Cu, Se and As atoms to have up to 4-fold coordinations would reach a limit at 32 at % Cu. The experimental limit to the glass-forming region occurs at about 31 at % Cu.[2]

Thus the structural model for the amorphous $Cu-As_2Se_3$ system consistent with the EXAFS results is that the Cu atoms are always 4-fold coordinated while the Se coordi-nation increases from 2 to 4-fold and the As coordination from 3 to approximately 4-fold as the Cu concentration increases from 5 to 25 at %. Accompanying the coordina-tion number change is necessarily the introduction of $Cu-Cu$, $As-Cu$ or $As-As$ bonds. The data was not sufficient to distinguish clearly between the backscattering amplitude of these similar atoms to determine the number of each kind of neighbor.

We note that these EXAFS results of coordination numbers around each specific kind of atom in these amorphous materials could not be determined by other structural techniques. Indeed, this represents the first direct measurement of the increased Se coordination in amorphous materials referred to in models of the conductivity in amor-phous semiconductors by Mott[11] and by Cohen, Fritzsche, and Ovshinsky.[12]

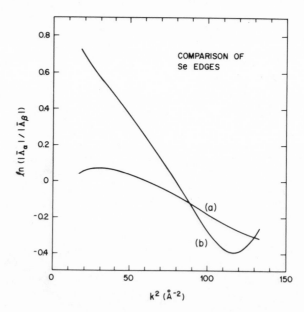

Fig. 3.   Comparison of nearest neighbor shells: $\ln(|\Lambda_\alpha|/|\Lambda_\beta|)$ versus k^2 for
(a) $\alpha = a-Cu_{.05}(As_{.4}Se_{.6})_{.95}$ $\beta = c-As_2Se_3$
(b) $\alpha = a-Cu_{.25}(As_{.4}Se_{.6})_{.75}$ $\beta = c-As_2Se_3$

c- and a-GeSe

Crystalline and amorphous *GeSe* (stoichiometric) will be considered briefly here as an example of the application of EXAFS to an amorphous system without dilute components. The structure of the $Ge-Se$ system has been studied extensively by many techniques. Previous x-ray RDF studies[13] could fit equally well two models for the nearest neighbor coordination in $a-GeSe$. Crystalline *GeSe* has each *Ge* atom 3-fold coordinated with *Se* atoms and each *Se* atom 3-fold coordinated with *Ge* atoms, rather than the typical covalent coordination. One model for $a-GeSe$ is this same 3-fold/3-fold coordination of the *Ge*/*Se* atoms. The second model has each *Ge* atom 4-fold coordinated by three *Se* and one *Ge* atoms and each *Se* atom 2-fold coordinated by two *Ge* atoms (as in covalent bonding). Figure 4 shows a comparison of the magnitudes of the direct Fourier transforms of the $k\chi(k)$ data of the *Ge* and *Se* absorption edges for both $c-$ and $a-GeSe$. The height and width of the first neighbor peak is a measure of the combined effect of the coordination number N and the disorder σ^2. For $c-GeSe$, the equivalence in peak magnitudes for *Ge* and *Se* edges is consistent with the 3-fold/3-fold model. On the other hand, the magnitudes of these peaks favors the 4-fold/2-fold model for $a-GeSe$ (note that the peak magnitude of the *Ge* edge is *ca* twice that of the *Se* edge), however, a more accurate determination of the coordination numbers and disorder must await accurate inverse Fourier transform analysis.

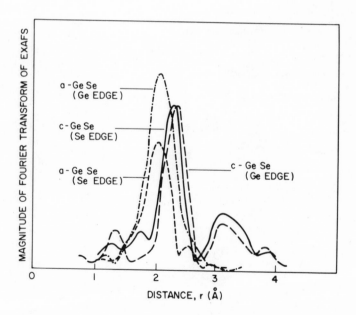

Fig. 4.   Comparison of Fourier transforms of EXAFS of *Ge* and *Se* Edges in $c-GeSe$ and $a-GeSe$.

In conclusion, the EXAFS analysis of these two amorphous systems demonstrates that EXAFS definitely is applicable to amorphous materials and that it indeed provides unique and useful structural information not available from other techniques. The most effective use of EXAFS is to choose amorphous materials problems with an awareness of the limitations of the method and to combine it with other structural information. Then, good data acquisition and careful data analysis can yield accurate determinations of coordination numbers, coordination distances, degree of thermal or structural disorder and possibly the identities of neighboring species.

REFERENCES

1. See, for example, E. A. Stern, D. E. Sayers, and F. W. Lytle, Phys. Rev. **B11,** 4836 (1975).

2. K. S. Liang, A. Bienenstock, and C. W. Bates, Phys. Rev. **B10,** 1528 (1974).

3. See the chapter by A. Bienenstock elsewhere in this book.

4. D. E. Sayers, E. A. Stern, and F. W. Lytle, Phys. Rev. Lett. **27,** 1204 (1971).

5. T. M. Hayes and S. H. Hunter, in *The Structure of Non-Crystalline Materials,* ed. P. H. Gaskell, Taylor & Francis, London, 1977, p. 69.

6. S. H. Hunter, A. Bienenstock, and T. M. Hayes, in *The Structure of Non-Crystalline Materials,* ed. P. H. Gaskell, Taylor & Francis, London, 1977, p. 73.

7. B. K. Teo and P. A. Lee, J. Am. Chem. Soc. **101,** 2815 (1979).

8. S. H. Hunter, Ph.D. Thesis, Stanford University, 1977.

9. F. W. Lytle, private communication.

10. A. Bienenstock and S. H. Hunter, unpublished.

11. N. F. Mott, Adv. Phys. **16,** 49 (1967).

12. M. H. Cohen, H. Fritzsche, and S. R. Ovshinsky, Phys. Rev. Lett. **22,** 1065 (1969).

13. A. Bienenstock, J. Non-Cryst. Solids **11,** 447 (1973) and references cited therein.

EXAFS OF DILUTE SYSTEMS: FLUORESCENCE DETECTION

J. B. Hastings

Physics Dept.
Brookhaven National Laboratory
Upton, New York 11973

Introduction

Since the first observations of the variation of the absorption coefficient for X-rays above the energy thresholds in the thirties until the early seventies, measurements and analysis of these variations were merely intended for the understanding of the underlying physics. Recently, with the understanding of the information available about the local atomic structure in the neighborhood of the absorbing species and the availability of high intensity synchrotron radiation sources, EXAFS has become a powerful structural tool. In the discussions that follow the details of the measurement for very dilute species will be presented. It is shown that for the more dilute systems the measurement of the emission rather than the direct absorption is a more favorable technique.

Fluorescence versus Transmission

The use of transmission EXAFS for concentrated systems is well established. When the systems under investigation become increasingly dilute in the absorbing atom there is a point where the signal to noise (S/N) ratio favors alternative techniques which measure signals characteristic of the absorbing species (x-ray fluorescence and Auger electron detection, for example). In the transmission mode, neglecting the statistics of the incident beam, the signal to noise ratio is given by[1]

$$S/N = 0.735\sqrt{I_o}\,\frac{\mu_x}{\mu_t}\,.$$

(1)

where I_o is the incident beam intensity and μ_x and μ_t are the absorption coefficient for the atom of interest and the total absorption coefficient of the sample, respectively. Both μ_x and μ_t are functions of the incident photon energy E and can be written as

$$\mu_x = N_x\,\sigma_x$$

(2)

and

$$\mu_t = \sum_i N_i\,\sigma_i$$

(3)

where N is the density and σ is the absorption cross section, and the sum over i is for all atoms in the sample.

The signal to noise ratio for the case of fluorescence detection has to include the efficiency for detecting the characteristic X-rays from the atom of interest and the scattered intensity of the incident photon and may be written as

$$S/N = \frac{I_f \epsilon_f}{\sqrt{I_f \epsilon_f + I_s \epsilon_s}} . \qquad (4)$$

where ϵ_f and ϵ_s are the detection efficiences for the fluorescent and scattered photons and I_f and I_s are the fluorescent and scattered intensities, respectively. I_f and I_s are given by

$$I_f = \frac{I_o \, \omega_k f_k (\Omega/4\pi) \cdot \mu_x(E)}{\mu_t(E) + \mu_t(E_f)} \left\{ 1 - \exp\left[- \left[\mu_t(E) + \mu_t(E_f) \right] d \right] \right\} \qquad (5)$$

and

$$I_s = \frac{I_o \mu_t^*(E)(\Omega/4\pi)}{2\mu_t(E)} \left\{ 1 - \exp(-2\mu_t(E)d) \right\} . \qquad (6)$$

Here ω_k is the probability of producing a fluorescent photon of energy E_f after the creation of the appropriate core hole by an incident photon of energy E. f_k is the probability of filling the core hole with a given electron which produces a specific fluorescent photon energy. Both μ_t and μ_x, which have been defined previously, are explicitly written as functions of the incident energy E and the fluorescent photon energy E_f. $\mu_t^*(E)$ is the total scattering cross section for the sample including both the coherent and incoherent contributions. Finally, Ω is the solid angle subtended by the detector and d is the sample thickness. The angular dependence of the scattering cross sections due to the plane polarization of the incident beam has not been included.

By equating the signal to noise ratios for transmission and fluorescence for a given sample one can solve for N_x/N_t to get the concentration at which these two techniques are comparable. Assuming that the detection system cannot discriminate between fluorescent and scattered photons, $\epsilon_f = \epsilon_s$, and that the fluorescent intensity equals the scattered intensity, $I_s = I_f$, then

$$\frac{\mu_x}{\mu_t} = \left[\frac{1}{1.47} \right]^2 f_k \omega_k (\Omega/4\pi) \qquad (7)$$

If we further assume that $\sigma_x \sim \sigma_t$, $\omega_k \sim 0.5$ (ω_k for Cu is $\sim .5$), $f_k \sim 1$ and the detector subtends about 1% of the 4π sterad, we obtain

$$N_x/N_t \sim 2000 \; ppm. \qquad (8)$$

This qualitative estimate yields a lower limit for the concentration of the transmission technique. Is there a lower limit for the fluorescent technique? To answer this question, we shall assume that we have a perfect detector such that $\epsilon_f = 1$ and $\epsilon_s = 0$. Then the signal to noise is given by

$$S/N = \frac{I_f}{\sqrt{I_f}} . \qquad (9)$$

For a thick sample, $d \to \infty$, I_f is given by

$$I_f = \frac{I_o \, N_x \, \sigma_k \omega_k f_k \, (\Omega/4\pi)}{\mu_t(E) + \mu_t(E_f)} .$$ (10)

The variation of the product $\sigma_k \omega_k f_k$ as a function of the atomic number Z for K_α fluorescent radiation is shown in Figure 1.

To get an estimate for I_f we shall evaluate I_f at threshold energy where $E \sim 1.1 E_f$ for the K edge. Assuming $\mu_t(E) \propto (1/E)^3$, we have

$$I_f \approx I_o \, \frac{N_x}{N_t} \, \frac{\sigma_k \omega_k f_k}{\sigma_t} \cdot 0.43 \cdot (\Omega/4\pi) .$$ (11)

For $(\Omega/4\pi) = 0.01$, $I_o = 1 \times 10^{11}$ photons/sec which is typical of current storage ring synchrotron x-ray sources and an impurity concentration of $N_x/N_t = 100$ ppm, I_f for an $Fe \, (Z = 26)$ or a $Mo \, (Z = 42)$ impurity in various hosts is given in Table I. For an impurity with a Z comparable to that of the host, the total signal would be on the order of 10^4 to 10^5 counts/second. The EXAFS, however, represents somewhere between one and ten percent of the total signal. A one percent EXAFS therefore requires typically 10^6 counts/data point. It can easily be seen that the lower limits on concentration will always be source limited with a perfect detector and for the example shown may be of the order of 10 to 100 ppm of a low Z impurity in a high Z host.

For thick samples the ratio of the scattered to fluorescent intensities can be calculated from Eq. 5 and 6, as $d \to \infty$

$$\frac{I_s}{I_f} = \frac{\mu^*(E)/2\mu_t(E)}{\omega_k f_k \mu_x(E)/(\mu_t(E) + \mu_t(E_f))}$$ (12)

Assuming that no edges exist between E and E_f then $\mu_t(E_f) \approx (E/E_f)^3 \mu_t(E)$ and at threshold energy $E \sim 1.1 \, E_f$, we have

$$\frac{I_s}{I_f} \approx 1.2 \, \frac{N_t \, \sigma_t}{N_x \, \sigma_x f_k \omega_k} .$$ (13)

With the value of $N_x/N_t = 1 \times 10^{-4}$ (100 ppm of the dilute specie) values of I_s/I_f have been calculated as a function of the atomic number of the host for several impurities. The results are shown in Figure 2. It is readily apparent that for all the impurities, I_s/I_f is typically 300 for hosts with atomic numbers similar to that of the impurity.

Detectors

At the heart of fluorescence EXAFS is the detection hardware. In this section three possibilities will be discussed: an integrating detector, a filter assembly[2] and a crystal analyzer,[3] with emphasis on the crystal analyzer. Again, the signal to noise ratio for fluorescence EXAFS is

$$S/N = \frac{I_f \epsilon_f}{\sqrt{I_f \epsilon_f + I_s \epsilon_s}}$$ (14)

Factoring out $\sqrt{I_f}$ yields

$$1/\sqrt{I_f} \cdot S/N = \frac{\epsilon_f}{\sqrt{\epsilon_f + (I_s/I_f)\epsilon_s}}$$ (15)

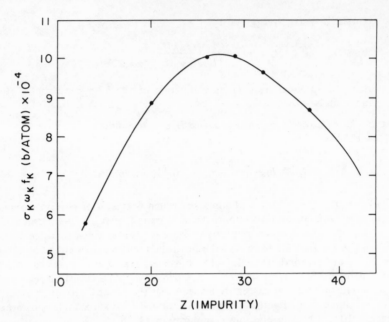

Fig. 1.   Plot of the variation of the product of photoelectric cross section for the K hole, σ_k, radiative transition probability, ω_k, and the probability of filling the K holes with L electrons, f_k, as a function of the atomic number Z of the impurity.

Table I: Fluorescent Intensities I_f (counts/sec) for Fe or Mo Impurities in C, Cu, and Nb Hosts (assuming $\Omega/4\pi \approx 0.01$ and $I_0 \approx 1 \times 10^{11}$ photons/sec).

Impurity	Host	I_f
Fe	*C*	3.8×10^6
Mo	*C*	1.2×10^8
Fe	*Cu*	5.6×10^4
Mo	*Cu*	8.8×10^4
Fe	*Nb*	1.3×10^4
Mo	*Nb*	2.6×10^4

For fixed incident intensity, the three types of detectors can be compared using Eq. 15. Assuming the following detection efficiencies

$$\epsilon_f = \epsilon_s = 1 \qquad \text{integrating}$$

$$\epsilon_f = 0.5, \ \epsilon_s = 0.05 \qquad \text{filter assembly}$$

$$\epsilon_f = 0.1, \ \epsilon_s = 0 \qquad \text{crystal analyzer.}$$

the values of $1/\sqrt{I_f} \cdot S/N$ can be calculated as a function of I_s/I_f for the three types of detectors using Eq. 15. The results are shown in Figure 3. It is apparent that the values of I_s/I_f for which the crystal analyzer is equivalent to the integrating detector and the filter assembly in terms of $1/\sqrt{I_f} \cdot S/N$ are 8 and 29, respectively. It should also be noted that the "normalized" signal-to-noise $1/\sqrt{I_f} \cdot S/N$ is independent of the solid angle. It is also emphasized that the crystal analyzer is advantageous for the most dilute systems, while the filter assembly is superior for the less dilute systems (N_x/N_t greater than 100 ppm).

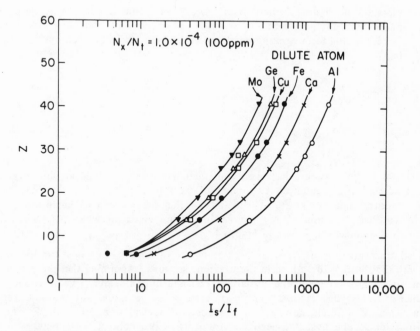

Fig. 2.   Plots of the atomic number Z of the host as a function of the ratio of scattered to fluorescent intensities I_s/I_f for 100 ppm concentration of several impurities.

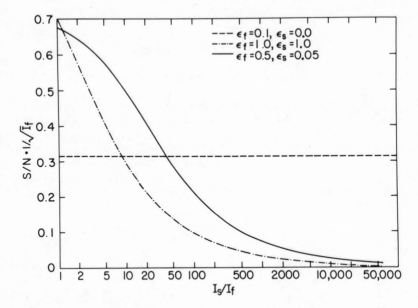

Fig. 3. Plots of the figure of merit $S/N \cdot 1/\sqrt{I_f}$ as a function of the scattered to fluorescent intensity ratio I_s/I_f for 3 choices of detector efficiencies corresponding to integrating detector, filter assembly and crystal analyzer (see text).

A crystal analyzer system has been constructed utilizing pyrolytic graphite crystals that can be held in a vacuum chuck to give a large solid angle and provide focusing to further improve signal to noise. The details of this scheme and its application to a study of internal oxidation of 75 ppm *Fe* in *Cu* are given elsewhere.[3] A schematic diagram of this detector is shown in Figure 4.

Finally the variation of N_x/N_t for a given choice of impurity and host can be calculated with the formulas described in the previous sections. Assuming that the detectors subtend an equal solid angle, the crystal analyzer and filter assembly give an equal performance at $I_s/I_f = 29$. Figures 5 and 6 are plots of the variations of impurity concentration versus Z (atomic number) of the host for several impurities and Z of the impurities for several hosts, respectively, for $I_s/I_f = 29$. Both of these Figures show that except for high Z impurities in low Z hosts, the concentrations of impurity are greater than 100 ppm and in many cases as large as 1% for $I_s/I_f = 29$.

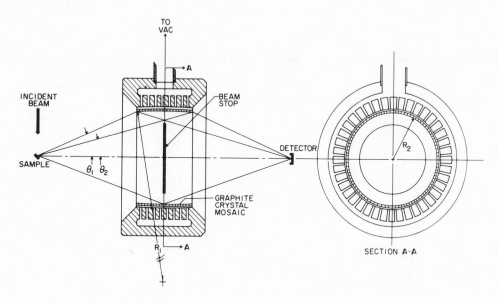

Fig. 4. Schematic diagram of a Rowland's Circle geometry crystal analyzer described in the text and in Ref. 3.

Summary

With the design and construction of dedicated storage rings, the development of EXAFS analysis, and advances in detection techniques, the investigations of the structure about dilute species are becoming more routine. With the aid of the calculations presented here the best techniques for the study of a specific system can be evaluated. In general for systems of biological interest where the Z of the dilute specie is large compared to the host the filter assembly is preferred. However, for metallurgical systems it may often be necessary to use a crystal analyzer.

ACKNOWLEDGEMENTS

I wish to thank Dr. B. M. Kincaid for useful discussions and Dr. P. Eisenberger for a continuing collaboration. Special thanks also go to my colleagues at the National Synchrotron Light Source, in particular Dr. M. L. Perlman. This work has been supported by DOE contract EY-76-C-02-0016.

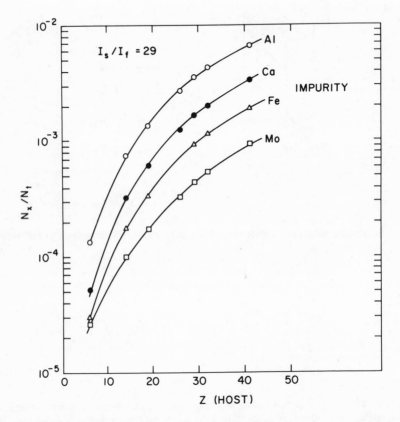

Fig. 5.   Variations of concentration N_x/N_t of some dilute systems as a function of the atomic number Z of the host for several impurities. I_s/I_f is fixed at 29 (see text).

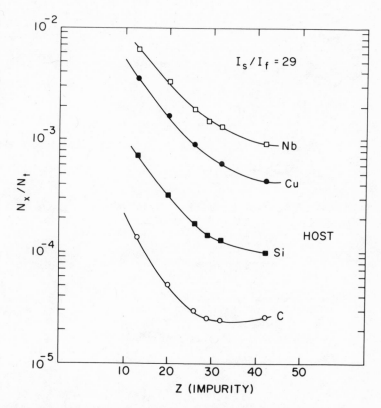

Fig. 6.   Variations of concentration N_x/N_t of some dilute systems as a function of the atomic number Z of the impurity for several hosts. I_s/I_f is fixed at 29 (see text).

REFERENCES

1. J. Jaklevic, J. A. Kirby, M. P. Klein, A. S. Robertson, G. S. Brown, and P. Eisenberger, Solid State Comm. **23,** 679 (1977).

2. E. A. Stern and S. M. Heald, Rev. Sci. Instr. **50,** 1579 (1979).

3. J. B. Hastings, P. Eisenberger, B. Lengeler, M. L. Perlman, Phys. Rev. Lett. **43,** 1807 (1979).

EXAFS STUDIES OF DILUTE IMPURITIES IN SOLIDS

Matthew Marcus

Bell Laboratories
Murray Hill, New Jersey 07974

Introduction

There are many interesting phenomena that happen in dilute solutions of impurities in solids. Such phenomena include "fast" diffusion, enhancement/dehancement effects of impurity concentration on host or impurity diffusion, and doping in semiconductors. Further, the isolated impurity in a crystal lattice provides a useful "test bench" for ideas about interatomic interactions in solids.

Most methods for studying the structure of solids respond to both the host and impurity atoms. Therefore, it is very difficult to perform structural investigations on systems with impurity contents below about 1%. What is needed is a technique which provides information about the environment of the impurity atoms alone.

The fluorescence EXAFS method may be used with very small impurity concentrations, and can provide information about the local geometry about the impurity atom. Also, EXAFS has the unique capability of providing information about the lattice vibrations around the impurity, and thus about the interatomic forces between host and impurity atoms.

In this chapter, several examples of the use of EXAFS in dilute impurity systems will be discussed. This review is not meant as a complete survey, since in such an expanding field, much work is in progress but not in print.

Impurities in Crystalline Solids

The first system to be discussed is that of Cu in Al. This system has several metastable phases in the range from 0-2.5 at. % Cu. One of these phases is the θ' structure, in which the Cu atom is surrounded only by Al atoms at known distances. This phase is thus an ideal model compound from which the phase and amplitude information can be extracted. This phase was used as a model by Lengeler and Eisenberger[1], and by Fontaine, Lagarde, Naudon, Raoux, and Spanjaard[2]. They both looked at the solid solution of Cu in fcc Al.

Lengeler and Eisenberger[1] found that the best fit to the data could be obtained by assuming the Cu atom to have 12 Al neighbors and 0.4 Cu neighbors. The inclusion of the Cu-Cu pairs improved the fit by a factor of two, which shows that the Cu-Cu pairing produces a marginally significant effect on the data. Since the Cu content was 0.5 at. %, the average Cu-Cu coordination would be 0.06 if random mixing applied. The actual value of 0.4 shows that there may be some Cu-Cu clustering. The Cu-Al distance was $(2.79\pm.03)$Å, to be compared with the Al-Al nearest-neighbor spacing in the Al lattice of 2.85Å. Thus, there is an apparent contraction of the Al lattice around the Cu atom.

Fontaine, et. al.[2] measured a distance of $(2.725\pm.02)$Å in a 2 at. % solution, with no consideration given to the possibility of Cu-Cu clustering. The discrepancy between the distance values of the two groups may be due to a large amount of clustering in the 2% solution as compared with the 0.5% solution, or due to a different analysis technique.

Fontaine, et. al.[2] also measured the EXAFS from Cu in Guinier-Preston (GP) zones. GP zones are thought to be monolayers of Cu sitting substitutionally on (100) planes, with the nearest Al (100) planes distorted so as to bring the Al atoms closer to the Cu layer than the "ideal" distance. That picture was confirmed by Lengeler and Eisenberger[1], who found that the Cu atoms are surrounded on the average by 2 Cu and 2 Al atoms in the zone plane at a distance of 2.86 Å, and 8 Al atoms out of the plane, at a distance of $(2.63\pm.02)$Å. The argument given for the mixed in-plane coordination, however, depends on precise amplitude transferability, which may not be valid[3].

We thus see that EXAFS may be used to detect the lattice distortion around an impurity, and to confirm models in which the impurity atom is in a fairly complex structure, such as a GP zone. More information could have been obtained from EXAFS measurements in the Al-Cu system by looking at higher shells, systematically varying the impurity concentration, and analyzing the GP zone data with methods more sophisticated than those used in Ref. 2.

The impurity concentration in the above investigations were sufficient to permit the use of transmission EXAFS. The study described below was carried out using fluorescence EXAFS with the aid of a graphite concentrator/discriminator described in the previous chapter by Hastings.

When an alloy containing small amounts of Fe in Cu is annealed in oxygen, its low-temperature resistivity becomes much less than that of an alloy of the same composition which has been heat-treated in hydrogen. For an Fe concentration of 75ppm, the resistivity-reduction ratio is about 1800. This effect is thought to be due to the aggregation of the Fe impurities into oxide precipitates. Hastings, et. al.[4] have studied a 75ppm Fe in Cu alloy in both oxidation states in order to determine in what form the Fe exists in oxidized and reduced states.

For the reduced alloy, they find that there are 12 nearest-neighbor Cu atoms around the Fe, at a distance of 2.54Å, while the ideal nearest-neighbor spacing is 2.56Å. Though there is no indication of nearest-neighbor Fe, it is not clear that Fe neighbors can be distinguished from Cu neighbors by EXAFS.

In the oxidized state, the Fe atoms seem to be neighbored by Cu and O atoms. The distances derived are different from those expected for FeO, Fe_3O_4, or $CuFe_2O_4$, which are the most plausible candidates for the Fe-rich phases (see discussion in Ref. 4). Thus, the EXAFS data leave one with a mystery, which has yet to be solved. It is possible that a heretofore-unknown phase has been discovered. Otherwise, the precipitates may have been distorted by the strains produced by the Cu matrix.

Certain bcc metals exhibit "anomalous" diffusion, in which the diffusivities of certain impurity atoms, as well as those of the host, show activation energies and pre-factors which are much lower than one would expect from vacancy-mediated diffusion. Such a system is Cu in Ti, in which Cu and Ti are "anomalous" diffusers in bcc (and probably hcp) Ti. A solution of 0.5 at. % Cu in hcp Ti was studied by Marcus[5] in order to determine the impurity sites as well as to understand the impurity-host interaction. The Cu-Ti distance was determined to be (2.875±.045)Å, while the Ti-Ti distance, averaged over the split first shell of the hcp structure, is 2.923Å. Thus, the Cu must be quite close to the center of a substitutional site, with little lattice distortion. The Cu coordination was 8.4, which given the possible amplitude transferability errors incurred by the data analysis, is not far from the nominal value of 12.

The EXAFS Debye-Waller factors (DWF's) provide information about the thermal fluctuation in the distance between the central Cu atom and the neighboring Ti atoms. The EXAFS DWF for Cu in Ti was about the same as that for Ti in Ti, indicating similar force constants for the Ti-Ti and Cu-Ti interactions.

Now, let us compare the Ti lattice, which supports "anomalous" diffusion, with the Cu lattice, which does not. The atomic masses and elastic moduli of both lattices are nearly equal, thus simplifying the analysis. The X-ray diffraction DWF, which samples the long- and short-wavelength lattice modes with equal weight, is about the same for the two pure metals. However, the EXAFS DWF, which samples only the short-wavelength modes, is larger by a factor of two for the Ti lattice as for the Cu lattice.

It therefore seems reasonable to suppose that the short-range forces in Ti are much less important than they are in Cu, relative to the long-range forces. It is thus possible that defects having no long-range strain-field, such as interstitial-vacancy pairs, may be more stable, relative to monovacancies, in the Ti lattice than in Cu. While this argument is still very speculative, it shows that the EXAFS DWF contains useful and important information. Also, the data show that a Cu atom in the Ti lattice is very much like a Ti atom in the same position. Thus, we see that Cu is an "anomalous" diffuser in Ti for the same reasons that Ti is itself an "anomalous" diffuser.

Impurities in Amorphous Solids

We now leave the realm of metal atoms in metals, and consider the problem of impurities in amorphous systems. Here, there is no hope that diffraction or channeling will provide any useful information about the impurity sites. Therefore, EXAFS, despite its limitations in disordered systems, can be one of the most useful probes of impurities in amorphous systems. In covalent systems, the first-neighbor distance may be fixed by chemical bonding as to make negligible the corrections due to disorder.

It is well known that amorphous Si-H alloys can be doped with small amounts of As, which change their conductivity by many orders of magnitude. Knights, et. al.[6], have done EXAFS measurements on As in Si-H, with As contents of 1-12 at. %. The Si-As distance was 2.38Å, which is quite close to the Si-Si distance of 2.35Å. This result shows that the As fits into the a-Si structure without causing large distortions. The coordination number for most of the alloys was between 2.5 and 2.8, as predicted by a model in which the hydrogen distributes itself at random among Si and As atoms. The hydrogen does not make any detectable contribution to the EXAFS and therefore can be ignored in EXAFS data analysis. The 1% As sample showed slightly greater coordination than would be predicted by the random model. This discrepancy may signal an incipient transition to a state in which the As is 4-coordinated to the Si. The evidence for this effect is not yet totally convincing. The 12% As sample showed no detectable As-As bonding, thus indicating that As-As bonds are less favored than Si-As bonds.

Another possible interpretation of the coordination is that most of the As is three-coordinated with Si, and that most of the H is bonded to the Si. The As-Si bond length in crystalline AsSi is about equal to the As-Si and Si-Si interatomic distances in the glass. It seems that the "natural" As-Si distance is close to the Si-Si length, thus allowing easy incorporation of As into the Si structure.

A further example in this context is that of 5 and 25 at. % Cu in amorphous As_2Se_3, which has been discussed in detail by Hunter in a previous chapter.

Conclusions

We have seen that EXAFS can provide some valuable and often unique information about the local coordination, geometry, and vibration near an impurity atom in a solid. The case of an isolated impurity in a crystal can be an ideal one for EXAFS study, due to the well-defined nature of the host crystal structure. With the new generation of wide-energy-range focussed and/or wiggler beamlines at dedicated synchrotron light sources, and with new detection technology now being developed, it will be possible to do measurements on solutions dilute enough to preclude any possibility of clustering. The data gathered in this region will contribute to our understanding of diffusion, alloying, doping, and other phenomena. As mentioned previously, EXAFS is sometimes the only technique that can yield direct information about the siting of impurities in amorphous materials. Systems such as the ones discussed above may be the ones in which the use of EXAFS is the most justified and the least dispensable.

REFERENCES

1. B. Lengeler, P. Eisenberger, Phys. Rev. **B21**,4507(1980).

2. A. Fontaine, P. Lagarde, A. Naudon, D. Rauox, D. Spanjaard, Phil. Mag. **40B**,17(1979).

3. P. Eisenberger, B. Lengeler, Phys. Rev. B **22**,3551(1980).

4. J. B. Hastings, P. Eisenberger, B. Lengeler, M.L. Perlman, Phys. Rev. Lett. **43**,1803(1979).

5. M. Marcus, to be published in Solid State Commun.

6. J. C. Knights, T. M. Hayes, J. C. Mikkelson, Jr., Phys. Rev. Lett. **39**,712(1977).

MATERIALS RESEARCH AT

STANFORD SYNCHROTRON RADIATION LABORATORY

Arthur Bienenstock

Stanford Synchrotron Radiation Laboratory
Stanford, California 94305

INTRODUCTION

In this paper, a few examples of materials research presently underway at the Stanford Synchrotron Radiation Laboratory are presented. These include x-ray diffraction techniques utilizing anomalous scattering, studies of surfaces utilizing photoemission, and x-ray small angle scattering. EXAFS, one of our major endeavors, is not included because virtually all such studies described in this volume were performed at SSRL, so that a description here would be redundant.

OVERVIEW OF SSRL

SSRL is a national users' laboratory supported by the National Science Foundation. No fees are charged for its general services and provision of photons, although payment may be required for special services. Potential users should consult SSRL Proposal Guidelines[1] for more information about gaining access to its experimental facilities.

Its synchrotron radiation source is the electron-positron storage ring, SPEAR, 50% of whose operating time is presently dedicated to the production of synchrotron radiation. During the other 50%, the synchrotron radiation production is parasitically dependent upon high energy experimentation. The ring is capable of storing electron currents at energies up to approximately 4.0 GeV with a resulting critical energy ϵ_c, of 11.2 GeV from its bending magnets. In the spring of 1979, a seven-pole, 18 kilogauss wiggler magnet providing synchrotron radiation with a critical energy of 19.2 KeV was installed. The resulting spectra are shown in Figure 1 as function of both photon and stored electron energies. One additional feature of the radiation that should be noted is that, when run in single bunch mode, the light pulse lengths range from 50 to approximately 200 picoseconds with a time between pulses of approximately 780 nanoseconds, making it an ideal source for many time-resolved studies.

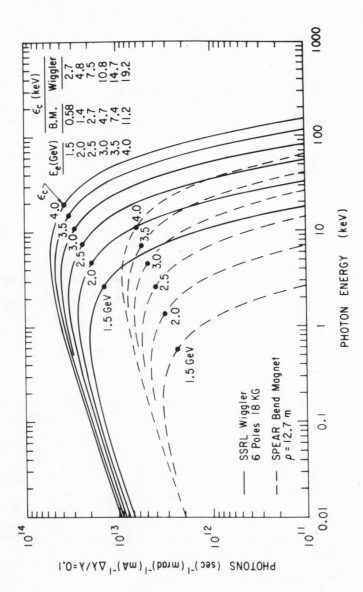

Fig. 1.   Calculated spectra from bending magnets (dashed lines) and existing wiggler (solid lines) at SSRL as a function of stored electron energy, E_e.

An overview of the Laboratory, showing its placement with respect to the storage ring, is presented in Figure 2. The experimental areas are housed in two large buildings adjacent to the ring plus one small experimental room mounted on top of the ring. The North Arc Building contains the original bending magnet Beam Lines I and II, and is presently being expanded to provide room for a new beam line which is to be illuminated by a ten-pole, 18 kilogauss wiggler magnet. (An undulator is also being prepared for utilization on either this beam line or Beam Line IV.)

The South Arc Building contains two beam lines. Beam Line III is devoted to soft x-ray and vacuum ultraviolet studies while Beam Line IV is the wiggler illuminated x-ray beam line. This building will also contain provisions for the preparation and maintenance of biological samples as well for the preparation and examination of x-ray lithographs and micrographs.

The characteristics of SSRL's 13 experimental stations, all of which may be used simultaneously, are presented in Table I. More detailed information is presented in the Proposal Guidelines, the article by Bienenstock[2] and the references contained therein.

ANOMALOUS SCATTERING

With the availability of expanded funding, more beam lines and dedicated time, SSRL has expanded its x-ray activities beyond the previously-dominant, experimentally simple, EXAFS experiments to more time and equipment consuming diffraction experiments. This is a trend which we may expect to continue as more experimental stations become available and we gain experience with the control and alignment of large diffractometers in essentially inaccessible x-ray beams. Among the most important reasons for the utilization of synchrotron radiation in diffraction experiments is the phenomenon of anomalous x-ray scattering. Far from an x-ray absorption edge, the x-ray atomic scattering factor, f, is reasonably well represented by the Fourier transform, f_o, of the electron density. There are, however, real, f', and imaginary, f'', shifts which are appreciable close to an absorption edge. That is.

$$f = f_o + f' + \mathrm{i}f'' \tag{1}$$

With the availability of tunable synchrotron radiation, several groups have been measuring f' and f'' close to absorption edges. Fukamachi et al.[3] have utilized the phenomenon of total x-ray reflection for this purpose on samples of copper. Phillips et al.[4] as well as Templeton et al.[5] have obtained these parameters by fitting the intensities of Bragg reflections measured with photon energies close to absorption edges for crystals of known structure to obtain values of f' and f'' for Cs, Pr, Sm, V and Co.

Finally, Fuoss et al.[6,7] have obtained values for Ge and Se in amorphous $GeSe$ through measurements of the x-ray absorption coefficient as a function of photon energy. These yield f'' almost directly, while f' is obtained from a dispersion relation. The results of this approach for Ge are shown in Figure 3.

It is apparent from this figure that quite large values of f' are obtained at photon energies just below the absorption edge. The maximum magnitude value of -13 is quite appreciable compared to the maximum value of f_o, 32, at $s(4\pi\sin\theta/\lambda) = 0$. With increasing s, f_o decreases monotonically, so that f' becomes an ever increasing fraction of the total scattering.

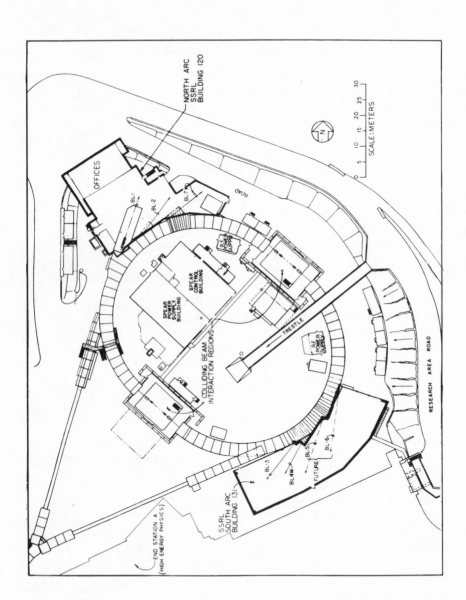

Fig. 2.   An overview of the electron-positron storage ring, SPEAR, which shows the SSRL North Arc and South Arc buildings and beam lines.

Table I

CHARACTERISTICS OF SSRL EXPERIMENTAL STATIONS

Beam Line I	Horizontal Angular Acceptance (Mrad)	Energy Range (eV)	Resolution	Approximate Spot size Hgt × Wdth (mm)	Remarks	Experiments (Examples)
I-1 (4°)	2.0	32-1000	$\Delta\lambda = .1-.2$ Å	1.0 × 2.0	DOUBLE FOCUS	PHOTOEMISSION PHOTOABSORPTION
I-2 (8°)	4.0	4-40	$\Delta\lambda = .2-6$ Å	1.0 × 3.0	HIGHLY POLARIZED DOUBLE FOCUS	PHOTOEMISSION REFLECTANCE SCATTERING FLUORESCENCE
I-3 (90° X-ray)	1.0	8400	~ 1 eV	1.0 × 15.0	HIGHLY POLARIZED FIXED WAVELENGTH	SMALL ANGLE SCATTERING
I-4 (Curved Crystal)	2.5	6000-9500	~ 8 eV	0.5 × 0.5	DOUBLE FOCUS	SMALL ANGLE SCATTERING
I-5 (EXAFS I)	1.0	3800-29300	$\Delta E/E \sim 10^{-4}$	2.0 × 20.0	RAPIDLY TUNABLE	EXAFS FLUORESCENCE SCATTERING
Beam Line II						
II-2	4.5	2800-8500	$\Delta E/E \sim 10^{-4}$	2.0 × 4.0	DOUBLE FOCUS RAPIDLY TUNABLE	DIFFRACTION SURFACE EXAFS
II-3	5.5	2800-9500	$\Delta E/E \sim 10^{-4}$	2.0 × 4.0	DOUBLE FOCUS RAPIDLY TUNABLE	EXAFS DIFFRACTION
II-4	1.0	3200-3000	WHITE RADIATION	4.0 × 15.0		X-RAY TOPOGRAPHY
Beam Line III						
III-1 (4°)	2.0	32-800	$\Delta\lambda = .05-2$ Å	1.0 × 1.0	DOUBLE FOCUS	PHOTOEMISSION ABSORPTION
*III-2 (18°)	3-5	4-150	?	?		GAS PHASE EXPERIMENTS
III-3 (2°)	8-10	800-4000	0.5-5 eV	1.0 × 2.5	ULTRA HIGH VACUUM	EXAFS PHOTOEMISSION
*III-4 (3°)	2.0	2-3000	WHITE RADIATION	?		MICROSCOPY LITHOGRAPHY
Beam Line IV (Wiggler)						
IV-2 UNFOCUSED	1.0	2800-45000	$\Delta E/E \sim 10^{-4}$	2.0 × 20.0	RAPIDLY TUNABLE	EXAFS FLUORESCENCE SCATTERING
FOCUSED	5.0	2800-10000	$\Delta E/E \sim 10^{-4}$	2.0 × 6.0	RAPIDLY TUNABLE	EXAFS FLUORESCENCE SCATTERING
IV-3	1.0	2800-50000	$\Delta E/E \sim 10^{-4}$	1.5 × 15.0	RAPIDLY TUNABLE	EXAFS
Beam Line VII (Wiggler)						
*VII-2	1.0	3200-4500	$\Delta E/E \sim 10^{-4}$	2.0 × 5.0	RAPIDLY TUNABLE	EXAFS FLUORESCENCE SCATTERING
90° UV (Lifetimes Port)	2.0	1-7	WHITE RADIATION	2.0 × .2	PULSED VISIBLE LIGHT	FLUORESCENCE LIFETIMES

*Under Construction   February 1980

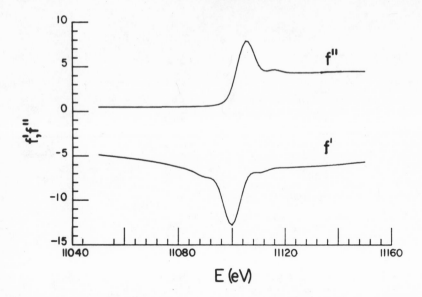

Fig. 3.   Anomalous scattering factors of *Ge* in amorphous *GeSe*, as measured by
Fuoss et al. (Reference 6).

The initial work described above has concentrated on determining values of f' and
f'' close to absorption edges in specific, important individual systems. These measurements are important because it was essentially impossible, without synchrotron radiation, to measure these quantities with high photon energy resolution close to the edge. It
should be pointed out, however, that we still lack information about the s-
dependence of f' and f'' close to the edge and possible variations of these quantities
with the chemical environment of the atomic species. The phenomenon of EXAFS as
well as, in some materials, white lines are clear indications that f'' varies with chemical
environment close to the edge. This, in turn, implies that f' must also vary with that
environment. The magnitudes of these effects are not, however, known yet.
Nevertheless, it has been possible to utilize these measured values in important experiments described below.

AMORPHOUS BINARY ALLOYS-PARTIAL PAIR DISTRIBUTION FUNCTIONS

As discussed by Bienenstock,[8], it is frequently impossible to obtain a unique structural model for an amorphous binary alloy from an x-ray, electron or neutron
diffraction radial distribution function (RDF). The RDF is defined by the Equation

$$D(r) = \frac{1}{N} \Sigma_i \Sigma_j K_i K_j \delta(r - r_{ij}) \tag{2}$$

where the summations run over all the atoms in the sample, N is the total number of atoms and r_{ij} is the distance between atoms i and j. The weighting factors, K_i, are the atomic numbers for an RDF derived from x-ray diffraction and the scattering cross section for a neutron diffraction function. The RDF has its peaks at common interatomic distances, while the areas under these peaks are related to the coorrdination numbers. Ambiguity in interpretation frequently arises because $A-A$, $A-B$ and $B-B$ pairs may contribute to any peak, so that a unique chemical specification cannot be made.

Recently, Fuoss[7] has shown that it will be possible, in a variety of alloys with the atomic numbers larger than roughly 25, to markedly reduce the ambiguity using anomalous scattering and synchrotron radiation.

The origin of this possibility is evident from examination of the Debye equation for the scattering from an isotropic array of atoms, which may be expressed as

$$I(s) = N_A |f_A|^2 \int_0^\infty D_{AA}(r) \sin(sr)/(sr) \, dr$$

$$+ N_A (f_A f_B^* + f_A^* f_B) \int_0^\infty D_{AB}(r) \sin(sr)/(sr) \, dr$$

$$+ N_B |f_B|^2 \int_0^\infty D_{BB}(r) \sin(sr)/(sr) \, dr. \tag{3}$$

Here, $D_{\alpha\beta}(r)$ is the partial distribution function defined by

$$D_{\alpha\beta}(r) = \frac{1}{N_\alpha} \sum_{i=1}^{N_\alpha} \sum_{j=1}^{N_\beta} \delta(r - r_{ij}), \tag{4}$$

while s is the scattering vector magnitude, $4\pi\sin\theta/\lambda$. An alternative form for Eq. 3 is

$$I(s) = N_A |f_A|^2 S_{AA}(s) + N_A (f_A f_B^* + f_A^* f_B) S_{AB}(s) + N_B |f_B|^2 S_{BB}(s). \tag{5}$$

Comparing Eqs. 3 and 5, it is evident that the partial distribution functions, $D_{\alpha\beta}(r)$, can be obtained from the partial structure factors, $S_{\alpha\beta}(s)$, by Fourier transformation. Knowledge of the individual $D_{\alpha\beta}(r)$ would allow considerable definition of the structure.

As suggested by Shevchik,[9] partial delineation of these quantities can be obtained through differentiation of Eq. 3 with respect to f_A', yielding

$$dI(s)/df_A' = 2N_A (f_{Ao} + f_A') S_{AA}(s) + 2N_A (f_{Bo} + f_B') S_{AB}(s) \tag{6}$$

Fourier transformation of this quantity then yields a quantity proportional to

$$D_A(r) = Z_A' D_{AA}(r) + Z_B' D_{AB}(r). \tag{7}$$

$D_A(r)$ provides information about the coordination of A atoms in the same manner as EXAFS.

Fuoss has shown that the derivative and $D_A(r)$ may be obtained quite successfully through measurements of $I(s)$ at two photon energies close to the absorption edge of atom A. In particular, he has obtained $D_{Ge}(r)$ for amorphous $GeSe$ and shown that the Ge is threefold coordinated.

As discussed by Fuoss, this approach provides information which is complementary to that which EXAFS yields. Because measurements can be made in the low-s region which is inaccessible with EXAFS, accurate coordination number and further-than-near neighbor information is obtained. On the other hand, because the photon energies are determined by the absorption edge position, the high range limited to $4\pi/\lambda$ photon is usually too restricted to provide the high resolution where EXAFS yields.

As proposed by Keating,[10] still further definition of the individual $S_{\alpha\beta}(s)$ and, consequently, the individual $D_{\alpha\beta}(r)$ can be obtained through measurements of $I(s)$ at three photon energies (one just below the absorption edge (E_A) of A, one just below E_B and one far from both). Eq. 5 yields three linearly independent (but poorly conditioned) equations for the three unknowns, $S_{\alpha\beta}(s)$.

This approach is analogous to that employed in neutron diffraction where it is occasionally possible to make three samples with three different isotopes having markedly different scattering factors. Until recently, it was not apparent[8] that the anomalous scattering factor changes would be sufficiently large to allow a reliable determination of the partial structure factors and pair distribution functions. The preliminary results of Fuoss et al.[6] indicate that they probably are, and yield the promise of studies of thin films and the many elements for which there are not available isotopes with sufficiently different scattering factors so that neutron diffraction can be employed. In comparing this approach with neutron diffraction, however, it should be noted that the neutron cross-sections are much better characterized than are the anomalous scattering coefficients and that the range of diffraction data in the anomalous scattering experiments are limited by the lowest energy absorption edge to be utilized in the experiment. In the neutron diffraction case there is not such limitation. On the other hand, neutron diffraction does require bulk samples and the availability of appropriate isotopes.

ANOMALOUS SCATTERING STUDIES OF PARTIALLY ORDERED SYSTEMS.

There are many systems containing two different metal atoms (denoted A and B) that do not have complete ordering of the A and B atoms on their respective crystallographic sites. Instead, a significant fraction of the B atoms sit on A sites and vice versa. It is frequently true that the A and B atoms in such systems are adjacent on the periodic table. As a result, it is extremely difficult to determine the state of order with x-rays because the scattering which delineates that state depends on $|f_A - f_B|^2$, which is rather small for atoms which differ in atomic number by only one.

This quantity may be made quite large, however, by tuning the x-ray photon energy to a value just below the absorption edge energy of, say, A. Close to E_A, $|f_A - f_B|^2 \approx (f_A')^2$, which is quite appreciable when the tunability of synchrotron radiation is exploited effectively.

Recently, for example, Yakel[11] has determined the cation site-occupation parameter in a single crystal of cobalt ferrite (nominally $CoFe_2O_4$) using x-ray photon energies just below the Fe and CoK absorption edges. As Yakel states, "In this spinel structure system, the O atoms approach a cubic close-packed arrangement in which 1/4 of the tetrahedral and 1/2 of the octahedral interstices are regularly occupied by cations. The value of the site-occupation parameter, X, that describes the distribution of A^{2+} and B^{3+} cations in these interstitial positions may vary from 0 (a normal spinel, all A^{2+} ions in tetrahedral sites to 1 (an inverted spinel, all A^{2+} ions in octahedral sites)." Yakel finds that intensity data taken with $Cu\ K_\alpha$ data yields an $X = 0.98$, but that nearly identical measures of agreement could be achieved for any value of X between 0.5 and 1.0. With synchrotron radiation intensity data taken near the Fe K edge, however, X was pinned down to a value of 0.83 ± 0.01.

As part of the work, Yakel measured f' and f'' for Fe near the Fe K edge and found that the results do not differ appreciably from predicted values in the range 6358 to 7105 eV, although the precision in estimates of f'' is poor at the lower energies.

ANOMALOUS SCATTERING IN SINGLE CRYSTAL STRUCTURE DETERMINATION

Anomalous scattering may also be used in regular single crystal structure determination. In x-ray diffraction experiments, one measures the intensity, which is proportional to $|F|^2$, the absolute value squared of the structure factor. In order, however, to obtain the electron density one must know both $|F|$ and its phase. This is the classical "phase problem".

In protein crystallography, it is common to solve the phase problem utilizing the multiple isomorphous replacement method of Green et al.[12] In this method, one measures the intensities from the native protein crystal and at least two isomorphous derivatives in which there is a substitution of heavy metals. From the changes in intensity, the phases can be determine under the assumption that the structures of the three crystals are identical, except for the heavy metal substitutions.

Anomalous scattering can be utilized to achieve the same sort of scattering intensity changes achieved through isomorphous replacement. That is, the x-ray wavelength, rather than the heavy metal, is changed to achieve three independent measurements of the intensity.

The anomalous scattering approach has the immense advantage that a single sample may be used, so that no assumptions need be made about the equivalence of the structure for the three chemically different samples. The anomalous scattering method also, of course, negates the need for the growth of three chemically different crystals which may, in itself, be impossible.

This approach has been analyzed extensively recently by Phillips and Hodgson[13] and is being pursued at SSRL by Hodgson and coworkers.

SMALL ANGLE X-RAY SCATTERING

Small angle x-ray scattering is commonly used in materials science to determine the populations, shapes and sizes of voids in vapor deposited materials and the topology of phase separation in amorphous materials. The high intensity and high natural collimation of synchrotron radiation makes it a natural choice for such studies.

At SSRL, extensive studies of phase separation in amorphous semiconductors and metallic alloys have been underway for approximately two years. These take advantage of the short exposure times necessary to study systems as a function of composition. That is, large number of samples can be studied because of the short exposure times whereas laboratory systems frequently require a day of exposure per sample.

From the very start of SSRL, however, there was considerable interest in using synchrotron radiation to study dynamic phenomena such as the changes in long muscle spacings during the muscle contraction and extension cycle. In such studies, the cyclic nature of the phenomenon being studied allows one to average data over many cycles.

More recently, interest has grown in the study of non-cyclic, kinetic materials science phenomena. In particular, Stephenson and coworkers are attempting to study the early stages of phase separation as a homogeneous melt is cooled into a region of sub-liquidus immiscibility. Similarly, Long and coworkers are attempting to study the kinetics of asphaltene precipitation associated with heavy crude oil refining.

PHOTOEMISSION STUDIES OF SURFACES AND INTERFACES

The tunability of synchrotron radiation has made photoemission a very effective tool for studying surfaces and interfaces. This surface sensitivity arise in two ways. The first is that the mean free path of photoexcited electrons in virtually all solids is about 4-6 Å when the energies of those electrons are approximately 80 to 200 eV above the Fermi energy. Consequently, virtually all those electrons in that energy range which reach the detector in the photoemission experiment originate very close to the surface. In addition, the photoionization cross-sections of many elements show distinct maxima and minima as a function of final electron state energy. As a result, one can tune the photons to either enhance strongly or suppress strongly the photoionization of these elements in order to study them specifically or to remove their influence so that other elements in the same energy region can be studied. For example, one may want to suppress the photoionization of a metal d level so that the photoionization of an absorbed layer dominates the photoemission spectrum.

As an example of the effectiveness of this approach in studying materials science problems, let me review some recent work of Spicer et al.[14] on the formation of Schottky barriers on III-V semiconductor surfaces.

In their initial studies of these materials, they showed that there were virtually no surface states in the gap. The surface atoms rearrange themselves so that the Group III atom gives up one electron to negatively ionize the Group V atom. The effect of the atomic rearrangement is to push the resulting anticipated donor and acceptor states into the conduction and valence bands, respectively. The absence of gap states made suspect, of course, the classical picture of Schottky barrier formation and led to further investigations.

In the next step of the study, very low concentrations of metals, in some cases, and oxygen in others were deposited onto freshly cleaved surfaces. It was found that the surface Fermi energy became pinned at extremely low compositions of the deposited species. Typically, these concentrations amounted to less than a tenth of a monolayer of a deposit of material. Again, the picture of barrier formation through the deposition of a metallic slab on a semiconductor surface seems inappropriate.

Next, they sought to determine whether the deposited atoms were present as islands or as isolated atoms at these low concentrations. In one study, the Au 5d emission was studied for Au on $GaSb$ with 30 eV photons, as a function of Au concentration. This photon energy was chosen to enhance Au photoejection compared to that of the Ga and Sb, and illustrates the utilization of tuning to increase "signal-to-noise". The splitting of the Au 5d electron levels at low Au concentrations indicated that the atoms were present in isolation, rather than forming islands. This observation again indicated the inadequacy of the "metallic slab" model for the barrier.

In order to obtain further understanding, photons of 120 eV energy were used to excite photoemission from $GaSb$ on which various thicknesses of Au had been deposited. Photons of this energy were chosen because the mean free path should be of the order of 5 Å. One of the most striking results of this study was the finding that there was little decrease in the Sb 4d emission even after more than a hundred monolayers of Au had been deposited. This indicated that the Sb is rising to the surface of the material so that the stoichiometry of the $GaSb$ at the $Au-GaSb$ interface was changing. This result was subsequently corroborated by Sputter-Auger-ESCA analysis.

With these and other observations, Spicer et al. were able to show that the defect states required for barrier formation were actually produced by the metal or oxygen being deposited on the semiconductor surface.

CONCLUSIONS

In this paper, I have treated a few examples of Materials Science research underway at SSRL. Still to be developed, I believe, are a series of experiments which will examine the kinetics of materials transformations. these include dynamics EXAFS utilizing white radiation, a dispersing crystal and a position sensitive detector, as well as dynamic small and large angle scattering using both position sensitive and energy dispersive detectors and dynamic topography. All of these techniques are being explored at various synchrotron radiation facilities around the world and should soon graduate from the developmental to the structural tool stage.

ACKNOWLEDGMENTS

The work described herein was supported, in part, by the National Science Foundations through support to the Stanford Synchrotron Radiation Laboratory (Contract DMR-77-27489) which is run in cooperation with the Stanford Linear Accelerator Center and the U.S. Department of Energy.

REFERENCES

1. Stanford Synchrotron Radiation Laboratory PROPOSAL GUIDELINES AND GENERAL INFORMATION, February, 1980.

2. A. Bienenstock, in Proceedings of the National Conference on Synchrotron Radiation Instrumentation, National Bureau of Standards, Gaithersburg, Maryland, June, 1979, Nucl. Instr. and Methods, **172,** 13 (1980).

3. T. Fukamachi, S. Hosoya, T. Kawamura, S. Hunter and Y. Nakano, Jap. J. Appl. Phys. **17,** Suppl. 17-2, 326 (1978).

4. J. C. Phillips, D. H. Templeton, L. K. Templeton and K. O. Hodgson, Science **201,** 257 (1978).

5. L. K. Templeton, D. H. Templeton and R. P. Phizackerley, J. Amer. Chem. Soc. **102,** 1185 (1980); D. H. Templeton and L. K. Templeton, Acta Cryst. A, in press; D. H. Templeton, L. K. Templeton, J. C. Phillips and K. O. Hodgson, Acta Cryst. A, in press.

6. P. H. Fuoss, W. K. Warburton and A. Bienenstock, J. Non-Cryst. Solids **35-36,** 1233 (1980).

7. P. Fuoss, Ph.D. Thesis, Stanford University, 1980, unpublished.

8. A. Bienenstock, in *The Structure of Non-Crystalline Materials,* Proceedings of the Symposium held in Cambridge, England, Sept. 20-23, 1976, ed. P. H. Gaskell, Taylor & Francis, London, 1977, p. 5.

9. N. Shevchik, Phil. Mag. **35,** 805, 1289 (1977).

10. D. Keating, J. Appl. Phys. **34,** 923 (1963).

11. H. Yakel, to be published in J. Phys. Chem. Solids.

12. D. W. Green, V. M. Ingram and M. F. Perutz, Proc. Roy. Soc. **225,** 287 (1954).

13. J. C. Phillips and K. O. Hodgson, Acta Cryst., in press.

14. See, e. g., W. E. Spicer, P. W. Chye, P. R. Skeath, C. Y. Su and I. Lindau, J. Vac. Sci. Technol. **16,** 1422 (1979) and references contained therein.

CORNELL HIGH ENERGY SYNCHROTRON SOURCE: CHESS

Boris W. Batterman

Applied and Engineering Physics
Cornell University
Ithaca, New York 14853

Synchrotron radiation is now generally accepted[1] as an important scientific tool for the study of matter in all its forms. There are many synchrotron sources throughout the world providing radiation over a considerable portion of the electromagnetic spectrum. The new facility, CHESS, at Cornell University provides a source of radiation primarily in the x-ray regime at unprecedented intensities. This continuous radiation can be several orders of magnitude more intense than the line spectrum of a good x-ray tube (and therefore in the range of seven orders of magnitude more intense than the continuous background radiation from a tube). In addition, the synchrotron source has the striking property that it is confined to a very narrow angular spread, which allows the experimenter to combine the twin properties of high intensity and intrinsic collimation to enhance dramatically the number of photons incident on even very small specimens. In a sense, comparing the "brightness" of a synchrotron source to that of a conventional x-ray tube is analogous to comparing a laser to a conventional light source.

Origin of Synchrotron Radiation

Synchrotron radiation is emitted when relativistic charged particles move in curved paths in magnetic fields. It is, in fact the centripetal acceleration of the particle which, according to electromagnetic theory, is essential to produce the radiation. In the CESR storage ring, electrons and positrons are circulated in an evacuated chamber with an average·radius of about 100 meters. The particles travel in single bunches close to the velocity of light in a counterrotating sense. The bunches of positrons and electrons, each about 4 centimeters long, intersect twice per revolution. When they collide, new elementary particles are created. It is primarily for the purpose of elucidating the physics of these new particles that CESR was constructed. CESR at its design energy of 8 billion electron volts (GeV) and 100 milliamperes of electrons and positrons will emit in synchrotron radiation about 1 megawatt of electromagnetic radiation. In general, the total synchrotron radiation production varies as the fourth power of the electron energy.

197

The radiation from the centripetally accelerating electron is emitted in a narrow cone around the instantaneous path of the charged particle. The beam is similar to one swept out by the headlight of a train moving in a circular track. It is emitted in a plane parallel to the orbital plane of the particles. The angular height perpendicular to the plane is determined by the particle's relativistic energy. For CESR this corresponds to a vertical divergence of only 13 arc seconds and it is this remarkably small divergence that makes the source so intensely bright in the optical sense.

Synchrotron Radiation Facilities

There are basically two types of machines that provide synchrotron radiation. Those such as CESR are built and operated primarily for high energy physics experiments, and a small synchrotron radiation laboratory is attached in what is often called the parasitic mode. Sources of the other type have as their sole purpose the provision of synchrotron radiation, and they are called dedicated sources. In the United States, Tantalus I at the University of Wisconsin and SURF II at the National Bureau of Standards are dedicated sources for low energy vacuum-ultraviolet light. The colliding beam facility SPEAR at Stanford University has associated with it a major synchrotron radiation laboratory (SSRL) operating in a parasitic mode. (In the coming year, up to 50 percent of the operating time of SPEAR will become dedicated for SSRL.) At Orsay, France, there is a dedicated low energy ring and a parasitic high energy machine. In Germany, DORIS, part of the DESY synchrotron complex at Hamburg, is a storage ring that has facilities for high energy synchrotron radiation. A testimony to the importance of synchrotron radiation is the growing number of storage rings being built solely for its production. In this country, a new machine, Aladdin, dedicated to the production of low energy photons, is under construction at the University of Wisconsin. At Brookhaven National Laboratory the dedicated National Synchrotron Light Source is under construction. This facility, when completed, will have two storage rings, one for high energy photons and another for ultraviolet light. In England, Germany, and Japan dedicated sources are also being built. It is clear that synchrotron radiation is riding a crest of interest throughout the world.

To get an idea of the synchrotron radiation spectrum obtainable from some selected facilities, we show in Fig. 1 the computed spectrum expected at maximum design energies from the Brookhaven source, the SPEAR storage ring at Stanford, and the completed CESR ring at Cornell. A parameter useful in describing the spectrum is the critical photon energy E_c. This is defined as the photon energy at the midpoint of the integrated flux of the spectrum (that is, there is as much total flux above E_c as there is below). For CHESS this critical energy is 35 keV (corresponding to a wavelength of 0.35 Å). The flux at the critical energy is very nearly at the maximum of the photon spectrum. In Fig. 1 the critical energies are indicated by vertical arrows; it is clear that the range above 30 keV is unique to the Cornell facility, CHESS.

Cornell High Energy Synchrotron Source

For the past 15 years, the Laboratory of Nuclear Studies at Cornell has operated a high energy electron synchrotron with a maximum electron energy of 12 GeV that has been used for studies of particle physics by physicists from all over the world. For nearly 2 years the laboratory has been in the process of converting the synchrotron to an electron-positron storage ring (CESR) with a maximum design energy of 8 GeV.

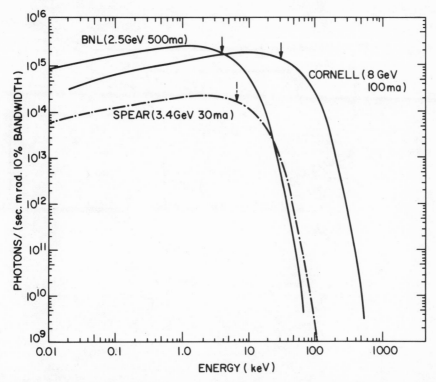

Fig. 1.   Spectrum of emitted synchrotron radiation as a function of photon energy.
Vertical arrows mark the critical energies. The curves shown are for Cornell
(CESR), Stanford (SPEAR), and Brookhaven (BNL).

Figure 2 shows the interaction area of the laboratory part of the CESR storage ring project. The main high energy physics experiments take place in the middle of the rectan-
gle marked "Cleo detector." To make room for this very large detector, the storage ring
(labeled "Colliding beam" in Fig. 2) had to be physically separated from the synchrotron, with the result that the two rings are separated by a straight section in CESR followed by a rather sharp bend into the interaction region. It is this sharp bend area
(which has a bend radius approximately one-third that of normal bending magnets) that
makes CESR a very potent source of synchrotron radiation.

The high E_c and therefore high x-ray energy content of the CHESS spectrum is a
direct result of the need to install high-bend magnets in the CESR storage ring.

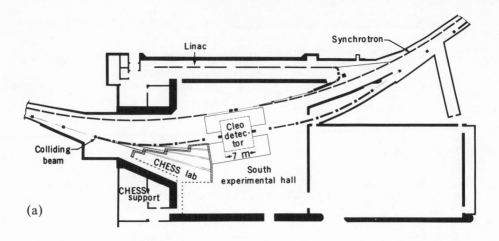

(a)

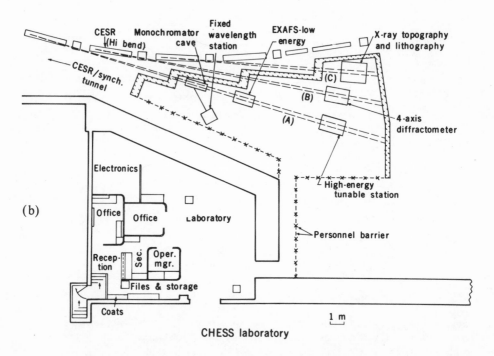

(b)

CHESS laboratory

Fig. 2. (a) A portion of the Wilson Synchrotron Laboratory operated by the Laboratory of Nuclear Studies at Cornell. The interaction area of the colliding beam project is at the center of the Cleo detector. Just before this, in the high curvature region of the colliding beam, can be seen the CHESS lines.

(b) Planned arrangement of the CHESS laboratory and the support area. The boxes show the approximate locations of the experimental stations.

Three independent beam lines have been activated at CHESS. Line A in the present configuration feeds a focused, fixed wave length station. We have obtained a beam at 1.2 Å and band width of about 5 eV, a flux of 10^{10} photons per second in a 1 mm^2 size. A second station on the A line is fed by a monochromator which is tunable from about 4 kilovolts up to 80 KeV. We have used this monochromator to study the gold K-edge. The monochromator feeds a fully-automated, 4-axis diffractometer and a mirror for the elimination of harmonics. The C line has two tunable monochromators which provide EXAFS capability in the range from 3 kilovolts up to approximately 50 KeV. The B line has the capability of providing a white beam for topography and dispersionless high pressure studies. At the present time, CHESS has the capability of operating five independent experimental stations.

In Fig. 3 the synchrotron radiation spectrum is shown as a function of beam energy for a given beam current. The vertical line represents a photon energy of 10 keV (wavelength, 1.2 Å). Over the anticipated operating energies of CESR, 4 to 8 GeV, the flux at this photon energy changes by a factor of only 7. Thus, regardless of operating energy, experiments using 1 Å radiation are always feasible regardless of the high energy physics requirements. From this point of view, users of moderately hard x-rays — that is, up to 10 keV — will not suffer from the parasitic use of synchrotron radiation operation.

The experimental stations at CHESS have problems unique to the high energy nature of CESR. The high energy tail of the synchrotron radiation spectrum makes shielding a little more demanding than in lower energy storage rings. In addition, monochromators that select needed wave lengths from the spectrum have been designed to minimize the high harmonic content that is always present because of the nature of the CHESS spectrum. We have developed mirror optics and tunable channel-cut crystals to virtually eliminate, when desired, the high harmonic content.

CHESS, supported by the National Science Foundation, is nationally available to users who need an x-ray source with energies from approximately 3 to 150 KeV. In the high energy end of the spectrum from approximately 30 Kilovolts on up, no other source under construction will offer comparable fluxes. Proposals for study at CHESS are currently being accepted and will be reviewed by a program committee if the demand exceeds the availability of the synchrotron radiation facility.

Applications: There are many unique characteristics of synchrotron radiation that make it especially applicable for studies of the structure of matter. Tunability allows one to investigate the scattering and absorption of x-rays in the vicinity of characteristic absorption edges of particular atoms of a specimen. For example, the variation of the absorption coefficient of x-rays in the vicinity of an absorption edge yields important localized structural information on the near-neighbor environment of the particular absorbing atom. This technique is called EXAFS (Extended X-ray Absorption Fine Structure). Tunability is an important new tool for the structural crystallographer. With the tunable source, the crystallographer can pick a wave length very close in energy to the absorption edge of a particular atom species in a molecule. Depending on the proximity of the chosen wave length to the edge, the scattering amplitude of that particular atom can be changed markedly. Then, by comparing x-ray crystallographic data taken with wave lengths near the edge to those with radiation far from an edge, the crystallographer can obtain information on the relative scattering phases of different

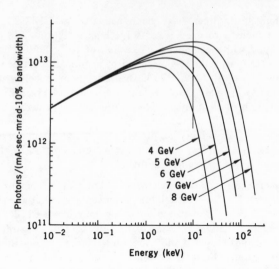

Fig. 3.   Spectrum of emitted radiation from CESR (see Fig. 1) as a function of beam
           energy, for a beam current of 1 mA.

atoms in the molecule and thereby gain additional information to help unravel the crystal structure. Heretofore, to get this information the crystallographer had to actually
replace particular atom types in the structure in a process called isomorphous substitution, but he had the difficulty of assuring that the subsequent arrangement of the
remaining atoms was unchanged. By using tunable synchrotron radiation, one can
simulate this isomorphous substitution without the necessity of tinkering with the structure of the molecule.

High energy tunable x-rays also offer exciting possibilities in x-radiography. It will
be possible to carry out x-radiography with high energy photons that are tuned very
close to the absorption edge of material injected into portions of the body. By taking
x-radiographs above and below an absorption edge, far greater selectivity can be
obtained. This is an area of x-ray dichromatography that has barely been explored.

The brightness and intrinsic collimation of a synchrotron x-ray source provide new impetus for the construction of x-ray imaging devices. It has already been demonstrated that a resolution of 100 Å can be obtained. The possibility now exists of carrying out in situ x-ray microscopy of protein matter in an aqueous environment with a resolution that may even exceed 100 Å. This has important biological applications and will complement structural electron microscopy.

X-ray diffraction by structures with widely spaced components usually involves scattering at extremely low angles. Low-angle x-ray scattering (LAS) which is an important tool for both biology and materials science requires a very highly collimated x-ray beam. Collimation of a conventional x-ray source, which emanates in all directions, is at the expense of a large reduction in intensity. Synchrotron radiation will have a strong impact in this field because the intrinsic high brightness and collimation can allow measurement of low-angle diffraction pattern in milliseconds, permitting real time experimental studies. An example of this is the diffractions studies of muscle contraction by Kulipanov and Skrinskii[2], in which the diffraction data were obtained in time intervals very small compared to the muscle contraction cycle time of 64 milliseconds.

Synchrotron radiation from a storage ring has an intrinsic pulsed time structure. For example, at CHESS, the light bursts come every 2.5 microseconds and have an intrinsic width of 0.13 nanosecond. These extremely short periodic bursts with a relatively long delay between pulses offers a wide range of possibilities in the study of time-resolved spectroscopy of excitation, emission, and decay of excited states in organic and inorganic molecules. The pulsed nature allows one to observe structural changes at controlled times after a specific stimulation, for example in a muscle or a nerve.

The experimental work at CHESS includes crystallography with anomalous dispersion, studies requiring the tuning of x-radiation across critical atomic levels, small angle scattering experiments, high pressure structures, x-ray topography, and a wide range of topics in materials science and biology.

What is exciting for us at Cornell, and for our colleagues nationally, is that a new and very intense source of synchrotron radiation is now available which will allow us to explore areas heretofore have not been accessible.

REFERENCES

1. H. Winick and A. Bienenstock, Annu. Rev. Nucl. Part. Sci. **28,** 33 (1978).

2. G. N. Kulipanov and A. N. Skrinskii, Sov. Phys. Usp. **20,** 559 (1977).

NATIONAL SYNCHROTRON LIGHT SOURCE (NSLS):

AN OPTIMIZED SOURCE FOR SYNCHROTRON RADIATION

J. B. Hastings

Physics Dept.
Brookhaven National Laboratory
Upton, New York 11973

Synchrotron radiation has a number of special characteristics that make electron storage rings and synchrotrons powerful tools for research with photons. Applications in both basic research and technology are possible and there are exciting prospects for the future with the design and construction of electron storage rings dedicated to synchrotron radiation production. Synchrotron radiation has a continuous spectrum ranging from the infrared to X-ray wavelengths. It is both intense and strongly polarized in the plane of the electron orbit. The light is emitted in pulses that are typically 1×10^{-9} sec long and have repetition frequencies of the order of 1×10^6 pulses per second. With the current designs, beam cross sections of 1 mm^2 or less can be achieved and the photons are sharply collimated in a narrow cone about the tangent to the orbit of the emitting particle.

Since the first sustained research program with synchrotron light began in 1961 art the National Bureau of Standards utilizing a 180 MeV electron ring, all the research programs world wide have used machines that have been designed for other purposes, high energy physics in particular. Realizing that with an optimized electron storage ring the future research with synchrotron light could be even brighter, a number of countries have proposed and started construction of dedicated facilities in the 1970's. In the sections that follow some of the considerations for an optimal design of a storage ring will be outlined and the choices of parameters for the National Synchrotron Light Source (NSLS) at Brookhaven National Laboratory (BNL) will be presented. Also the policy for utilization of NSLS beam lines will be described for both the 0.7 GeV (VUV) storage ring and the 2.5 GeV (X-ray) storage ring.

Synchrotron Source Optimization

The considerations of source parameters and source optics for synchrotron radiation research were discussed in detail in the classic report by G. K. Green.[1] A more complete discussion of sources and the characteristics of synchrotron light can be found in Ref. 2. This section relies heavily on Ref. 3 which contains more pertinent information. The single most important parameter that describes the optical quality of the source is its brightness and minimum emittances are desirable in order to achieve the optimum source brightness. This can be done by the proper choice of the storage ring magnet lattice that has maximum radiation damping of the transverse betatron oscillations. This choice of lattice would not be the same as the optimal design for a e^+e^- colliding beam facility. The dipole fields can also be increased in the magnet lattice to increase the photon flux and the critical energy of the photon spectrum for a given electron energy.

Following Green,[1] the central brightness can be expressed as

$$B = N_K(0,\lambda)/2\pi\sigma_x\sigma_y \tag{1}$$

$$\text{with } x = x' = y = y' = 0$$

where x, x', y, y' are components of the four dimensional phase space of the source. σ_x and σ_y are the transverse dimensions of the electron source. $N_K(0,\lambda)$ is the intensity of photons per unit band pass (with the multipler K and the wavelength λ), solid angle and time. This expression is correct in the limit where the radiation opening angle, σ_r, is large compared to the angular variations in the electron beam trajectories $\sigma_r > \sigma'_x$, $\sigma_r > \sigma'_y$ where

$$\sigma_r \approx \frac{0.565}{\gamma}\left[\frac{\lambda}{\lambda_c}\right]^{.425}. \tag{2}$$

Here γ is the ratio of the electron's energy to its rest mass, λ is the wavelength of the photon and λ_c is the critical wavelength of the storage ring.

In general the NSLS designs are consistent with these restrictions. Noting that the transverse dimensions of the source are related to the vertical and horizontal emittances $\sigma_y = \sqrt{\epsilon_y\beta_y}$ and $\sigma_x = \sqrt{\epsilon_x\beta_x}$, the maximum brightness is obtained for minimum emittances for a given amplitude functions β_y and β_x, respectively. The vertical emittances ϵ_y is driven by a coupling of the radial and horizontal motion of the electrons. In practical storage ring designs this coupling can be held to typically 0.1 so that $\epsilon_y \sim 10^{-2}\epsilon_x$.

An approximate expression for ϵ_x has been developed by van Steenbergen[3] following Sands[4]

$$\epsilon_x \approx C_q\gamma^2[(\oint Hds)/2\pi\rho^2]. \tag{3}$$

Here C_q is a constant and ρ is the radius of curvature of the electron orbit. H is a function of the local dispersion of the electron beam and the line integral is evaluated around the storage ring. Therefore, the quantity in brackets is only a function of the magnetic lattice and has been evaluated for a number of storage rings.[3] Minimizing this quantity (0.055 for the 2.5 GeV ring at NSLS vs. 0.28 for the Photon Factory, the Japanese synchrotron light source, also at 2.5 GeV) gives the highest source brightness. Recalling the expression for the local spatial source size, the amplitude function β

should be a minimum at the source locations. This has also been achieved at the NSLS.

With a given magnet lattice the other fundamental parameter of choice is the radio frequency of the accelerating system to replace the energy loss due to synchrotron radiation. In Ref. 3 a detailed discussion was given for the choice of low, typically 50 MHz, versus high, typically 300 MHz, operation. Basically, the low frequency system is better for high beam current, with less problems from ion collisions in the vacuum envelope. On the other hand, the higher frequency provides shorter pulses better suited to timing experiments but will perhaps give lower beam current. The NSLS has gone with the low frequency system trying to optimize the brightness but foregoing some capabilities for timing experiments.

The last point in source design are considerations for the inclusion of special structures, namely, wigglers and undulators. In the case of wigglers, which are used to produce a harder spectrum extending to higher energy regimes, the source is a superposition of intensities from separate poles of the magnet. Hence, maximum source brightness is achieved with minimum values of β_x and β_y at the location of the Wiggler magnets. Also a minimum value, preferably zero, for the local dispersion function η is desired to keep the σ_x value small since more correctly

$$\sigma_x^2 = (\beta_x \epsilon_x + \eta^2 \sigma_e^2 / E^2) . \tag{4}$$

Here σ_e is the standard deviation in electron energy about E, the mean electron energy. Also with $\eta = 0$, there is no adverse effect on the source emittance; in fact, the installation of wigglers in straight sections with zero dispersion at NSLS would reduce the source emittance by 0.6 (see Ref. 3).

For undulators[5] the requirements on β_x and β_y are different. The undulator is a many pole magnet that produces a coherent superposition of radiation from each "pole", giving rise to an extremely bright, quasi-monochromatic source of radiation. For the functioning of these devices β_x and β_y must be modest so that σ'_x and σ'_y of the electron orbit are small.[6] This capability has been built into the 0.7 GeV storage ring at NSLS.

Experimental Utilization and Beam Lines at NSLS

In order to best match experimental equipment to the synchrotron source taking into account the above considerations two storage rings are in construction at the NSLS. Their principal design parameters are given in Table I.

The typical photon spectra from these rings are shown in Figure 1. Also shown in Figure 1 is the spectrum for a wiggler magnet, 6 Tesla field, to provide radiation at shorter wavelengths. Note the critical wavelength, the half power point in the spectrum, is reduced from the arc source value, 2.5 Å, to 0.5 Å resulting from the increase from 1.2T to 6T field.

A plan view of the NSLS facility is shown in Figure 2. As can be seen from this Figure there are possibilities for 16 beam ports on the 0.7 GeV storage ring and 28 beam ports on the 2.5 GeV storage ring. All of these beam lines will ultimately be instrumented in one of two ways, either by the NSLS as a facility line open to the general public or by a participating research team (PRT) which provides up to 100% of the funding in exchange for a maximum of three-fourths usage. The remaining quarter

Table I. Design Parameters for the X-ray and VUV Storage Rings at NSLS

	X-ray Source	VUV Source
Current, energy (A, GeV)	0.5; 2.5	1.0; 0.7
Circumference (m)	170	51
λ_c (Å)	2.5 (arc) 0.5 (wiggler)	31.6 (arc)
Emittance, ϵ_x (m-rad)	8×10^{-8}	9×10^{-8}
SR Power (kW)	300 (5 wigglers)	12
Source $4\sigma_x$, $4\sigma_y$ (mm²⁾	0.5 × 1.5 (arc) 0.1 × 0.9 (wiggler)	0.4 × 1.2 (arc)

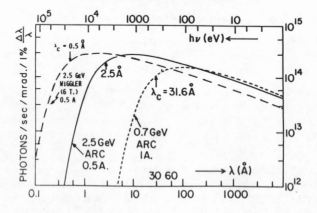

Fig. 1   Photon Spectra for NSLS. Plotted is log intensity per sec per mrad. per 1% band pass vs. log wavelength.

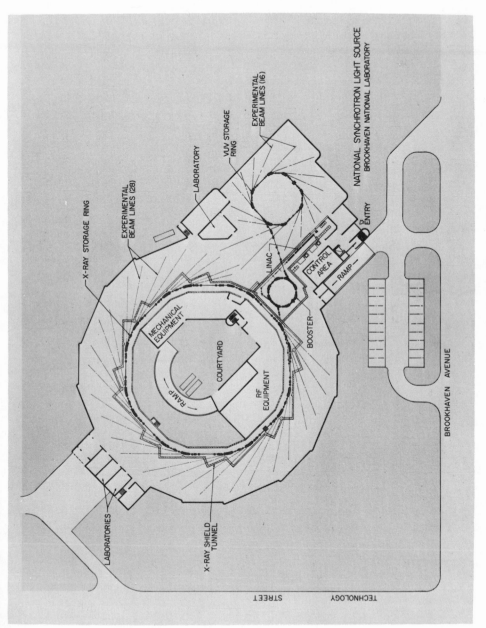

Fig. 2   Planned view of the NSLS facilities.

time utilization will be available to outside users with cooperation from the PRT. Thus the details of the experimental equipment that will be available to the general user is diverse and in many cases are still in the planning stage. Tables II and III comprise lists of most of the proposed experimental capabilities for the VUV (0.7 GeV) and X-ray (2.5 GeV) storage rings, respectively. Some of these lines are being constructed by PRT's and others, marked with asterisks, in whole or in part by the NSLS. It should also be mentioned that for some experiments, they may be several beam lines available. More details of the facility beam lines are contained in Refs. 7 and 8 for the VUV and X-ray storage rings, respectively.

Table II. Proposed VUV Beam Lines

*   HIGH RES. ARPES,[a] XPS[b] (35-1800 Å)
*   MEDIUM RES. ARPES, XPS, SEXAFS[c] (8-1200 Å)
*   SPECTROSCOPY — PHOTOCHEMISTRY
*   DYNAMICAL SPECTROSCOPY
*   BIOPHYSICAL SPECTROSCOPY
     IR SPECTROSCOPY
     ARPES, XPS
*   TEST LINE
*   FREE ELECTRON LASER

  * Constructed in whole or in part by NSLS.

[a]ARPES: Angle-Resolved Photoelectron Spectroscopy

[b]XPS: X-Ray Photoelectron Spectroscopy

[c]SEXAFS: Surface Extended X-ray Absorption Fine Structure.

Table III. Proposed X-Ray Beam Lines

*   SMALL ANGLE SCATTERING
*   EXAFS,[a] SEXAFS,[b] XPS[c]
*   X-RAY SCATTERING
*   TOPOGRAPHY
*   POWDER DIFFRACTION
     HIGH RESOLUTION SCATTERING
     DIFFUSE SCATTERING
     FLUORESCENCE ANALYSIS
     ATOMIC PHYSICS
*   WIGGLER
*   TEST LINE

  * Constructed in whole or in part by NSLS.

[a]EXAFS: Extended X-Ray Absorption Fine Structure

[b]SEXAFS: Surface EXAFS

[c]XPS: X-ray Photo Electron Spectroscopy

In summary, the NSLS is constructing two optimized storage rings, one operating at 0.7 GeV with a critical wavelength of 31.6 Å and the other at 2.5 GeV with a critical wavelength of 2.5 Å. Both of these storage rings will be instrumented by a combination of NSLS facility beam lines and beam lines provided by participating research teams.

ACKNOWLEDGEMENTS

The ongoing collaboration with W. C. Thomlinson, M. Howells, G. Williams, S. Krinsky, and A. van Steenbergen as well as other members of the NSLS staff is greatly appreciated. This work has been supported by DOE contract EY-76-C-02-0016.

REFERENCES

1. G. K. Green, BNL Report #50522 (1976).

2. S. Krinsky, M. L. Perlman, R. E. Waton, BNL Report #27678 (1980).

3. A. van Steenbergen, BNL Report #27654 (1979).

4. M. Sands, SLAC Report #121 (1970).

5. H. Motz, et. al. J. Appl. Phys. **24,** 826 (1953).

6. S. Krinsky, Nucli Instr. Methods, **172,** 73 (1980).

7. M. Howells, G. P. Williams, W. McKinney, BNL #26122 (1979).

8. J. B. Hastings, W. Thomlinson, BNL #26269, (1979).

ELECTRON ENERGY LOSS SPECTROSCOPY FOR EXTENDED FINE

STRUCTURE STUDIES — AN INTRODUCTION

David C. Joy

Bell Laboratories
Murray Hill, New Jersey   07974

INTRODUCTION

The spectroscopy of inelastically scattered fast electrons as a technique for structural and chemical studies of solids has been recognized since the pioneering work of Hillier and Baker.[1] A determination of the change in momentum of the incident electron, or separate measurements of the loss in energy and the angle through which the electron is scattered, provides a description of the scattering dynamics which can be interpreted macroscopically as a function of the complex dielectric coefficient of the material, or microscopically in terms of the chemical composition and physical state of the sample. Superimposed on the ionization "edges", which contain the significant chemical information in the spectrum, are fine structural features; details around the ionization energy which reflect the density of states available to the ionized electron, and extended fine structure modulations after the edge. The spectrum thus contains a complete chemical, physical and electronic description of the sample.

Practical Electron Spectroscopy

The obvious promise of this technique was not immediately fulfilled because of the practical problems associated with the needs for a high vacuum and a stable, high-intensity, high energy electron source. As these requirements became technologically easier to obtain, electron energy loss spectroscopy developed along two different routes. One was the special purpose electron scattering device optimized solely for spectroscopy. Typically[2] such instruments consist of an electron gun, operated at a low potential and temperature to ensure a minimum energy spread, followed by an accelerator stage to bring the beam to 100 or 200 keV. This electron beam is collimated by a simple electron optical system to illuminate an area of the sample some 50 μm to 1 mm in diameter. After transmission through the specimen, held at ultra-high vacuum to eliminate contamination, the scattered beam is decelerated to the original gun potential before passing into the energy analyzer. This configuration eliminates uncertainties due to drifts in the accelerating voltage, and it allows the spectrometer to be operated at a low energy, an arrangement which allows a high energy resolution ($<$0.2 eV) to be obtained with a relatively simple design. The disadvantages inherent in this approach

213

are the large volume of material sampled, and the poor collection efficiency of the spectrometer resulting from the optical magnification which occurs during deceleration.

The alternative, and later, line of development relied on the electron microscope. In their current form, transmission electron microscopes are able to image at, or close to, atomic resolution using either a fixed, wide ($1~\mu$m) collimated beam, or a convergent, finely focussed (1 nm) scanning beam. As an extension to these capabilities, other interactions induced by the electron beam can be used to obtain microanalytical information with a spatial resolution approaching that of the imaging mode. In particular an electron spectrometer can be added to the microscope column to collect the energy loss spectrum.[3]

By comparison with an electron scattering machine, an electron microscope can put a high electron flux into a very small selected area with the added benefit that the chosen area can be characterized by imaging and electron diffraction. Because of practical difficulties the spectrometer is invariably operated at the final accelerating potential of the gun, and a more sophisticated analyzer design is therefore required to achieve adequate energy resolution. However the post-specimen lenses of the microscope make it possible to optimally couple the spectrometer to the sample, giving both a high collection efficiency and sufficient (1 eV) energy resolution even with analyzers of modest specification. Under most practical conditions the energy resolution in this configuration is governed by the noise and stability of the high voltage supply and the natural energy width of the electron source.

Experimental Considerations

Extended fine structure modulations have been observed in energy loss spectra recorded with both types of instrumentation.[4,5,6] As shown in the following chapters, the theoretical performance of an electron source system for EXAFS studies should be comparable to, and often better than, a synchrotron-ring based photon system in terms of the achievable counting rate. In practice the two approaches have different strengths and weaknesses with the result that the particular details of the experiment planned will usually dictate whether a photon or electron technique should be used.

Electron spectrometer systems are most efficient for the studies of ionization edges between about 50 eV loss (the Li K-edge) and 2000 eV (above the Si K-edge), corresponding to the range over which photon systems are least efficient because of the high absorption of soft X-rays. On the other hand the electron ionization cross-sections fall away as E^4, where E is the energy loss, and consequently studies of heavy elements ($Z > 15$) are only possible by using the L- or M-edges. This restriction to low energy edges limits the energy loss range over which the EXAFS modulations can be followed because of the probability of encountering an interfering edge from another element. Since the smallest k vector value, k_{min} that can be used is about 4Å^{-1}, the resolution of the Fourier transform

$$\Delta R = \frac{\pi}{2(k_{max} - k_{min})} \text{ Å}$$

is limited to about 0.2Å, and in many other situations could be substantially worse (e.g. an organic compound containing C, N, O has edges separated by only 110 eV and 130 eV respectively).

The edges of interest in the energy loss spectrum ride on a background comprised of both characterized and non-characteristic scattering events. Particularly for minor elemental components in a complex system the peak to background ratio is usually small (i.e. less than unity) and so the quality of the data obtained will be governed by the accuracy with which this background can be modelled and stripped from the data. This probably limits the usefulness of the electron EXAFS technique to compounds in which the element of interest is at least 10 atomic percent of the irradiated volume.[7] It is clear that such a consideration indicates than an electron microscope-based system, which can focus the illumination onto very small chosen areas of the sample, has the advantage in cases where the sample is inhomogeneous since the volume fraction of the element of interest can be maximized in this way.

Because of the strong interaction between electrons and solids, the total inelastic mean free path is small (500-1000Å at 100 keV) and consequently plural scattering artefacts are evident in almost any spectrum obtained from a specimen of practicable thickness. Such effects (e.g. plasmon replication) are most serious immediately after the edge onset, and they set a limit to the minimum k-vector usable in the transform. While plural scattering can be reduced by increasing the accelerating voltage, the benefit is not great and for realistic operating energies (50-200 keV) sample thickness must not exceed 1000Å. This requirement is again easier to satisfy in an electron microscope based system, where the analyzed volume need only be 100-1000Å in diameter, than in an electron scattering machine, which needs a sample constant in thickness across a region tens of microns in size.

Finally it must be noted that electron induced contamination and radiation-damage of the specimen set limits to the counting statistics that can be obtained. While contamination can certainly be eliminated with proper care, radiation damage is a fundamental limitation and the significance of this is discussed in the chapter by Isaacson. The best hope for future progress lies in the development of parallel, rather than sequential, recording spectrometers.

SUMMARY

While the prospects for extended fine structure studies in electron energy loss spectroscopy are bright, the problems discussed above require solution before results, comparable to those now attainable from photon-sources, become routine.

Of the studies so far reported only that of Kincaid et al[4] using an electron scattering machine has data of a statistical quality comparable to that usually employed in photon-studies. While the results in that case demonstrate that the results of high quality can be obtained, the experimental conditions were sufficiently restrictive to limit the usefulness of that approach to cases where a particular strength of the electron technique, such as the ease with which data can be obtained for high q values, is required. The other studies published in this volume and elsewhere[5,6] which use an electron microscope have data which is of a significantly lower statistical accuracy but which was obtained under relatively straightforward experimental conditions. It is in this area of the electron technique that the greatest advances can be expected as more expertise in optimizing the instrumentation and sample preparation is obtained. For the electron microscopist the EXAFS technique in energy loss spectroscopy offers for the first time a way of effectively studying structures which are non-periodic at an atomic scale.

The X-ray and electron techniques are in many respects complementary rather than competitive and must be used accordingly although economic factors and questions of convenience are weighted heavily in favor of electron systems. The powerful combination of electron microscopy and electron spectroscopy should make it feasible to extend fine structure studies to systems too small for the more conventional technique and to add the dimension of spatial resolution to studies of inhomogeneous compounds.

References

1. J. Hillier and R. F. Baker, J. Appl. Phys. **15,** 663, (144).

2. A. E. Meixner, M. Schluder, P. M. Platzman and G. S. Brown, Phys. Rev. **B17,** 686, (1978).

3. D. C. Joy and D. M. Maher, J. Microscopy **114,** 117, (1978).

4. B. M. Kincaid, A. E. Meixner and P. M. Platzman, Phys. Rev. Lett. **40,** 1296, (1978).

5. R. D. Leapman and V. E. Cosslett, J. Phys. **D9,** L29, (1976).

6. P. E. Batson and A. J. Craven, Phys. Rev. Lett., **42,** 893, (1979).

7. D. C. Joy and D. M. Maher, Ultramicroscopy **5,** 333 (1980).

EXTENDED CORE EDGE FINE STRUCTURE

IN ELECTRON ENERGY LOSS SPECTRA

R. D. Leapman, L. A. Grunes, P. L. Fejes and J. Silcox

School of Applied and Engineering Physics
Cornell University
Ithaca, New York   14853

1. INTRODUCTION

Inner shell electrons can be excited not only by the absorption of X-rays but also by the inelastic scattering of fast electrons. The similarity between these two different probes suggests that experiments normally carried out with X-rays may be performed with electrons. This is the case for the Extended X-ray Absorption Fine Structure technique (EXAFS) developed by Sayers, Stern and Lytle.[1-5] In this chapter we shall discuss the equivalent use of electron energy loss spectroscopy (EELS) to give information about local atomic environments.

Recent experiments have demonstrated that spectra from thin films can be obtained with enough signal to make Extended X-ray-edge Energy Loss Fine Structure analysis (EXELFS) possible.[6-10] It has also been pointed out that the counting rates achievable with electron scattering can be competitive with those from synchrotron radiation experiments,[9,11] provided the core edges lie below about 3 keV, ie for the lower atomic number elements. The electron technique has a further advantage in that it can be carried out on a microscopic region by forming a narrow probe on the sample.[10]

However, the extension of the EXAFS technique to lower energies and various processes specific to electron scattering, create a number of difficulties in recording and analyzing data. The purpose here is to discuss these by means of examples and to try to make some assessment of the potential for the technique. It should be understood that the results we present are still preliminary as little systematic work has been reported for optimizing the experiments. We can therefore anticipate that substantial improvements to the current state-of-the-art will be made in the future.

2. COMPARISON OF EELS WITH X-RAY ABSORPTION

Cross Sections

The success of the EXELFS technique depends on there being enough signal to perform the type of EXAFS analysis possible with synchrotron radiation. Also it is important to ascertain whether the electron scattering technique has advantages over X-ray absorption. These questions have been addressed by Isaacson and Utlaut[11] in a comprehensive comparison of electron and photon beams and also by Kincaid et al.[9] We therefore do not go into detail here but concentrate on a few basic aspects where the two techniques differ since they are central to the present subject.

Two fundamental quantities are involved: the total flux of electrons or photons incident on the sample and the cross sections for inner shell excitation with each type of probe.

For X-ray absorption, the incident photon flux per unit area per unit photon energy $\frac{dJ}{dE}$ is the relevant quantity, while the cross sections $\sigma_{abs}(E)$ is simply related to the absorption coefficient $\mu(E)$ by,

$$\mu(E) = n\,\sigma_{abs}(E)$$

Here n is the number density of atoms in the sample. The count rate of photons transmitted through the sample per unit photon energy is,

$$\frac{dN}{dE} = A\frac{dJ}{dE}\,\exp(-\mu t) \tag{1}$$

where A is the area of sample exposed to the beam and t is its thickness.

The cross section per atom for excitation of an inner shell electron from an initial state $|i>$ to a final continuum state $|f>$ by absorption of a photon with energy E can be written as,[12]

$$\sigma_{abs}(E) = \frac{4\pi^2 e^2}{\hbar c}\,E\,\left|<f|\hat{\underline{e}}\cdot\underline{r}|i>\right|^2 \tag{2}$$

Here the final state $|f>$ is normalized per unit energy and $\hat{\underline{e}}$ is the polarization vector of the electric field. It is the interference between the outgoing and backscattered parts of the photoelectron final state that gives rise to the EXAFS oscillations in the cross sections.

Let us now consider electron scattering. In this case the incident energy and the total flux J is fixed since a nearly monochromatic electron beam falls on the sample. The count rate of scattered electrons which have been analyzed by a spectrometer and have lost energy E through core level excitation, is given in terms of the cross section per unit energy loss, $\frac{d\,\sigma_e(E)}{dE}$

$$\frac{dN}{dE} = AJnt\frac{d\,\sigma_e(E)}{dE} \tag{3}$$

For core level excitation by fast electrons with energy of order 100 keV, the cross section can be derived from the Born Approximation.[13] It is found not only to be a function of the energy transfer E, but also of the momentum transfer q. The same initial and final states are involved in the transition process but the operator in the matrix element is q-dependent.

Expressed in doubly-differential form the cross section is,[13,14]

$$\frac{d^2\sigma_e(E,q)}{dEdq} = \frac{8\pi e^4}{\hbar^2 v^2}\,\frac{1}{q^3}\left|<f|\exp(i\underline{q}\cdot\underline{r})|i>\right|^2 \tag{4}$$

From conservation of energy and momentum, q is related to the scattering angle θ by,

$$\frac{\hbar^2 q^2}{2mE_o} = \theta^2 + \theta_E^2 \tag{5}$$

where $\theta_E = E/2E_0$ with E_0 the incident electron energy and v its velocity. In the limit $q \ll \dfrac{1}{r_c}$, where r_c is the core orbital radius, the matrix element in eq. 4 reduces to the same dipole form as in eq. 2 with $\hat{\varepsilon}_q$ a unit vector along q in place of $\hat{\varepsilon}$. After integration over scattering angles up to θ_{max} we obtain,

$$\frac{d\sigma_e(E)}{dE} = \frac{4\pi e^4}{\hbar^2 v^2} \ln\left[1 + \left(\frac{\theta_{max}}{\theta_E}\right)^2\right] \left|<f\,|\hat{\varepsilon}_q\cdot\underline{r}|\,i>\right|^2 \tag{6}$$

The correspondence of the matrix elements in eq. 2 and 6 give the X-ray absorption and electron energy loss extended fine structure similar shapes.[14] Dipole selection rules, $\Delta l = \pm 1$, apply in both cases and the final state therefore has p-symmetry for the excitation of a K-shell. However despite the similar shape, the energy dependences of the two cross sections are different. The factor E in eq. 2 and the logarithmic term in eq. 6 make the electron scattering cross sections decrease relative to the X-ray absorption cross section with increasing energy.

Isaacson and Utlaut[11] have derived a 'figure of merit' corresponding to the ratio of fluxes per unit area on the sample for electron and photon probes multiplied by the ratio of their two cross sections. Typical values of parameters pertaining to electron and synchrotron sources were used. Results suggested that electron scattering should be competitive with photon absorption up to several keV of energy loss. These authors found that the figure of merit can be as high as 10^3 or 10^4 when a field emission source is used and when the electron spectrometer has a high angular collection efficiency. This estimation was made in the context of a microscopic scale and sensitivities were calculated for photon or electron fluxes incident in unit area on the sample. The ratio of total fluxes depends on the probe size. For synchrotron sources this is often a few $(mm)^2$, although the possibility of focusing to smaller dimensions with zone plates has been discussed.[11] For electrons high current densities (eg. 10^6 Amps cm^{-2}) are only possible when the probe size is small. Thus a field emission source can produce a 20 nA beam current in a 15Å diameter probe.[11] When the spot size is about $1(mm)^2$ the current density is several orders of magnitude lower, even for specially designed instruments with sources as bright as electron beam welders.

In practice the count rate for electron scattering becomes too small to obtain sufficient signal-to-noise ratio above a certain energy which is dependent on the precise limitations of the spectrometer collection efficiency and the beam brightness. Experience has so far shown[9,10] that above 2 or 3 keV it is difficult to attain the required statistics while at lower energy losses the cross section is more favorable. Later we shall estimate the expected count rates under typical experimental conditions in an electron beam instrument capable of forming a microscopic probe.

Other Considerations

In contrast to the electron scattering technique, X-ray absorption data is most easily obtained at energies above 2 or 3 keV. This is partly due to problems with optics;[15,16] gratings required to monochromatize the soft X-rays tend to suffer from contamination and impurity wavelengths occur in the monochromatized beam due to higher order

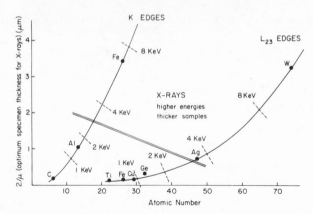

Fig. 1.   Optimum sample thickness $(2/\mu)$ for X-ray absorption experiments plotted
as a function of atomic number for K and L_{23} edges. Energies of the edges
are also indicated.

reflections. More importantly perhaps, difficulties in sample preparation should be considered. The incident photon beam is attenuated by the term $\exp(-\mu t)$ in eq. 1 from
which we obtain an optimum specimen thickness of $2/\mu$ for maximum signal-to-noise
ratio. It is crucial that the sample thickness t does not depart significantly from this.
Fig. 1 shows the approximate value of $2/\mu$ as a function of atomic number and threshold energy for K and L_{23} edges. The absorption coefficients were obtained from data
of Hubbell.[17] For a K-edge at about 2 keV, around silicon, the thickness should be
about 1 μm while the carbon K edge at 285 eV requires a thickness of only 2000Å. In
the case of the L_{23} edges at these energies, optimum thicknesses are even lower. Not
only must the samples be thin in this energy range but they must also be uniform over
a dimension of the beam size which may typically be a few $(mm)^2$. Thickness variations can give rise to distortions in the extended fine structure and introduce inaccuracies into analysis.[18] Sometimes it is possible to make thin uniform films by vacuum eva-
poration for example but in general many materials are not so readily prepared. Considerable difficulties can therefore exist in obtaining X-ray absorption data from lower
atomic number elements.

Electron scattering experiments also require thin samples. In fact the sample thickness is limited by plural scattering effects[8,19] which we shall discuss in more detail later.
Mean free paths for inelastic scattering by valence electron excitation (mainly
plasmons) are only about 500Å at 100 keV incident energy. We note that mean free
paths for the core excitations themselves are larger, often several μm in the energy
range of interest.

The main advantage with electrons is that they can be focused into a small probe so
that we only require sample uniformity on a scale of a few μm or possibly less. Instrumentation provided by the transmission electron microscope enables areas of thin film
to be imaged and other supplementary information to be recorded simultaneously such
as electron diffraction patterns which can help characterize the sample.[10] Combination
of the electron microscope with an electron spectrometer allows us to measure the
extended fine structure from a microarea. Moreover specimens can be prepared with
standard techniques applied to transmission electron microscopy.

3. EXELFS ANALYSIS

Basic Formula

Extended fine structure in the energy loss spectrum may be analyzed in the same way as EXAFS using the formulation of Sayers, Stern and Lytle.[1-5] The oscillatory part of the energy-differential cross section arising from interference between outgoing and backscattered parts of the final state is given by,

$$\chi(k) = \frac{1}{k} \sum_j \frac{N_j}{r_j^2} f_i(k) \exp[-2\sigma_j^2 k^2 - 2r_j/\lambda(k)]\sin(2kr_j + \phi_j(k)) \tag{7}$$

Here k is the wavenumber of the ejected electron which can be expressed as,

$$k = [2m(E-E_x)]^{1/2}/\hbar \tag{8}$$

where, $(E-E_x)$ is the energy above the core threshold at E_x. Summation is over the j different coordination shells each containing N_j atoms surrounding the excited atom. The backscattering amplitude function $f_j(k)$ depends on the type of atom. The term $\exp(-2\sigma_j^2 k^2)$ is a temperature factor taking into account thermal vibrations in the sample and $\lambda(k)$ is the range of the ejected electron, also a function of its kinetic energy.

The phase shift $\phi_j(k)$ has two parts, one due to the potential of the excited central atom and the other due to the backscattering atom. The phase shift experienced by the ejected electron in the field of the excited atom depends on the angular momentum of the initial state through the dipole selection rules. For K edges the $l=1$ phase shift is the relevant quantity. For L_{23} edges both the $l = 0$ and $l = 2$ phase shifts are involved. Electron-atom scattering theory[20] developed by Lee and Beni[21] has been used by Teo and Lee[22] to calculate phase shifts for K and L_{23} edges of nearly half the elements in the periodic table. These may be employed to analyze EXELFS in low atomic number materials where little other data exists.

The sum $\phi_T(k)$ of the phase shifts $\phi_c(k)$ and $\phi_b(k)$ due to central and backscattering atoms respectively may be parameterized by a power series. In the examples discussed below we keep only the term linear in k since we are not concerned here with high accuracy. For K edges the phase shift can be expressed as,

$$\phi_T(k) = \phi_c^{l=1}(k) + \phi_b(k) - \pi = a + bk + ... \tag{9}$$

For L_{23} edges the $l = 2$ part of the central atom phase shift is most important and the total differs by π compared with that for K-edges,[22]

$$\phi_T(k) = \phi_c^{l=2}(k) + \phi_b(k) \tag{10}$$

Let us briefly consider the validity of the EXELFS formulation (eq. 7) for large collection angles. The momentum transfer is then allowed to become large so that the dipole approximation no longer holds. Optically forbidden transitions are included and for K-shell excitation the ejected electron can have symmetry other than p. Phase shifts and therefore the fine structure might be affected. However, except at very high energies above threshold[14] and for typical spectrometer collection angles (about 10^{-2} rads at 100 keV incident energy), the largest contribution to the spectrum is still from optically allowed transitions. The formula above is therefore expected to apply under most conditions.

Steps in Data Reduction

In the analysis of our data we have used the following procedures some of which will be elaborated on in the examples below:

1. Subtraction of the background intensity preceding the edge by at least squares fit to an inverse power law,[23] $I \; \alpha E^{-r}$ where E is the energy loss and r is a constant approximately equal to 3 or 4 which may be found empirically.

2. Careful removal of the background after the edge by fitting the intensity to a low order polynomial, either third or fifth order.

3. Change the energy of the final state above threshold to a wavenumber scale according to eq. 8; $\hbar k$ then represents the momentum of the electron ejected from the core level.

4. Multiply by k^2 or k^3 to take account of attenuation by the backscattering amplitude function, $f_j(k)$.

5. Truncate the intensity modulations at k_{min} and k_{max}. The lower limit is chosen to be $\sim 2\text{Å}^{-1}$ below which data is not useful since solid state effects predominate. Here the range of the ejected electron is large and band structure of the solid influences the spectrum. The upper limit is set by the energy at which noise begins to dominate or where another edge occurs. In the examples below we choose $k_{max} \sim 10\text{Å}^{-1}$. The limited range of k available to carry out analysis is one of the main problems associated with the electron scattering techniques. It has the effect of limiting the resolution and accuracy in the radial distribution function.

6. The magnitude of the Fourier Transform is now calculated to give the radial distribution function, $F(r_j) = N_j/r_j^2$. For example this computation can be carried out on a minicomputer using a 1024 point Fast Fourier transform (FFT).

7. Finally half the linear part of the phase shift is estimated from the data of Teo and Lee[22] and applied to each pair of central and backscattering atoms. These values are added on appropriately to the peaks in the radial distribution function. Some assumption is of course involved as to the type of atom in each coordination shell.

We ignore the temperature and range factors in eq. 7. With such an approach we expect to be able to obtain near neighbor distances but only to deduce approximate information, if any, about coordination numbers.

4. COUNTING RATES AND INSTRUMENTATION

Previous Work

Extended fine structure above the core edges consists of small intensity modulations typically ranging in amplitude from 10% near threshold to less than 1% a few hundred eV higher in energy. Statistics are therefore required to be at the 1% signal-to-noise level and preferably 0.1%. This corresponds to between 10^4 and 10^6 counts per channel or perhaps even more if the background intensity preceding an edge is large. When we consider the relatively small probability of exciting the core electrons it is clear that optimized instrumentation is extremely important. Such factors have been crucial in the development of the X-ray absorption technique too. Here the comparatively long counting times required for conventional X-ray sources have been drastically reduced in recent years by the advent of synchrotron sources.[24]

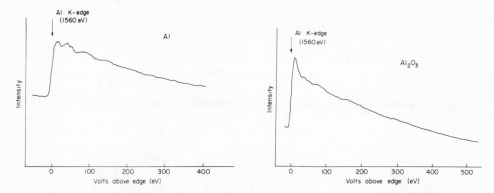

Fig. 2.   Extended fine structure above the aluminum K edge in Al and Al_2O_3 [Ref. 7]. Spectra were recorded from 200Å films at 60 keV incident energy and from an area about 0.5 μm in diameter.

We now consider existing EXELFS measurements, keeping in mind the count rates which have been achieved. Highest signals can be obtained with instruments that are specially constructed for electron scattering experiments. Ritsko et al.[6] used a machine which accelerated a monochromatized beam of slow electrons to impinge on the sample and then decelerated then to analyze their energy. Energy loss spectra were recorded from thin aluminum films and these authors demonstrated the occurrence of extended fine structures above the aluminum L_{23} edge at 73 eV, by comparing experiment with calculations.

Kincaid et al.[9] have measured the carbon K edge in graphite using scattering of 200 keV electrons in a similar instrument to that described above, and carried out the first detailed analysis of extended fine structure with EELS. These experiments were performed with a high incident electron flux of about 1 μA or 10^{13} electrons per sec. over an area about 1 mm across. Kincaid[9] obtained about 10^6 counts in a 0.5 eV channel at the carbon K edge with a few hours recording time for the entire spectrum. This was despite a limited collection efficiency of the spectrometer since a small angular acceptance was chosen. Nevertheless such a count rate is competitive with the present state-of-the-art achievable with synchrotron radiation in this energy range.

As was mentioned earlier the total probe current available in an electron microscope is less than $\sim 1\mu A$ so we cannot expect the statistics with this type of instrumentation to rival those potentially available with specially designed instrumentation. Extended fine structure was first recorded in a conventional transmission electron microscope (CTEM) by Leapman and Cosslett[7,8] using a magnetic sector spectrometer. In these measurements the aluminum K edge at 1560 eV was obtained from thin film samples of Al and Al_2O_3 from areas about 0.5 μm in diameter. These data are shown in Fig. 2 and are consistent with nearest neighbor spacings in the metal and oxide of 2.84Å and 1.90Å respectively. It was pointed out by these authors that the X-ray and electron techniques are complementary and can be applied to different types of samples.

Further EXELFS measurements were made on graphite in the CTEM by Rossouw,[25] and it was shown that carbon K-edge data agreed with calculations of the extended fine structure.

Isaacson and Utlaut,[11] and Batson and Craven[10] have discussed the high currents which can be focussed into small probes in the scanning transmission electron microscope (STEM) equipped with a field emission gun. Currents as high as 10 nA can be focused into probes only a few nm in diameter and the flux of electrons may be even greater if the beam diameter exceeds 100Å. Batson and Craven[10] have obtained EXELFS data from very small microareas using a commercial STEM. The spectrum in Fig. 3(a) was recorded by these authors from a thin 100Å graphite sample using a 15Å probe diameter at ~ 80 keV incident energy. Fig. 3(b) shows the residual EXELFS intensity multiplied by k^2 and displayed on a wave number scale. After transforming the data, the radial distribution function in Fig. 3(c) was obtained. Two peaks rise above the noise, the first of which Batson and Craven[10] attribute to the 1.4Å nearest neighbor spacing and the second to a composite of the 2.5Å and 2.8Å distances. It is noted that the 0.6 mrad collection angle used in this experiment to achieve adequate energy resolution, gave less than 1% efficiency and this might be improved substantially. Batson et al.[10,26] also detected differences in EXELFS from carbon films evaporated onto mica and potassium chloride substrates. They were able to compare results with diffraction data recorded from the same specimen areas and preliminary measurements indicate interesting differences.[26] Such work could be significant in the study of disordered materials.

Finally, Johnson and Csillag[27] have obtained extended fine structure data from thin graphite and aluminum films on a microscopic scale. These experiments were carried out in a CTEM with a scanning attachment. Radial distribution functions were extracted from the spectra by EXAFS - type analysis and near neighbor spacings agreed with the expected values.

Statistics

We have already seen that EXELFS measurements are feasible in the context of electron microscopes. Let us now estimate the counting rates achievable with an optimized system by making use of eq. 3 for dN/dE and eq. 6 for the cross sections.

We shall consider a STEM and choose typical parameters found in such an instrument, which is shown schematically in Fig. 4. We take a 100 keV beam of electrons, forming a probe on the specimen with a diameter somewhere between 10Å and 100Å, and a total probe current of 10 nA (a conservative estimate). We assume a magnetic sector spectrometer capable of 1 eV energy resolution and designed to accept electrons scattered inside a semi-angle of 10 mrads (Fig. 4).

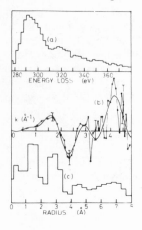

Fig. 3.   (a) Extended fine structure above the carbon K edge in a 100Å graphite film recorded in a STEM by Batson and Craven [Ref. 10] with a probe diameter of 15Å.
(b) Residual EXELFS intensity multiplied by k^2.
(c) Fourier Transform of data showing graphite radial distribution function.

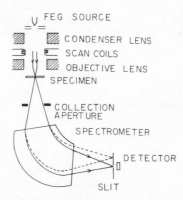

Fig. 4.   Schematic diagram of STEM showing magnetic sector spectrometer.

Consider a sample 200Å thick consisting of amorphous carbon. The K-shell excitation cross section just above threshold (285 eV) can be determined from calculated oscillator strength data,[28] and from eq. 6. We find

$$\frac{d\sigma_e}{dE} \sim 10^{-6}\text{Å}^2 eV^{-1} atom^{-1}$$

Putting this into eq. 3 gives a counting rate

$$\frac{dN}{dE} \sim 10^6 eV^{-1} sec^{-1}$$

Consider next a specimen of chromium of similar thickness (200Å) and we are interested in the intensity just above the L_{23} edge at 575 eV. In this case oscillator strength data[28] gives a cross section

$$\frac{d\sigma_e}{dE} \sim 0.5 \times 10^{-6}\text{Å}^2 eV^{-1} atom^{-1}$$

and the corresponding counting rate is

$$\frac{dN}{dE} \sim 5 \times 10^5 eV^{-1} sec^{-1}$$

Both the carbon and chromium spectra therefore require counting times of about 1 sec per eV channel at these core edges. However because of the decreasing cross sections above threshold, the average time per channel would need to be somewhat larger.

Some Aspects of Instrumentation

The estimate of counting time given above, assumes serial data collection. That is, the spectrum is scanned past a single detector by ramping for example the magnetic prism current. If a parallel detection system were used incorporating a position sensitive device like a diode array,[29] the entire spectrum could be recorded in a few seconds rather than several minutes. Although no fully optimal scheme has yet been devised for parallel detection of the energy loss spectrum, work is in progress in this direction.

Another important requirement is that the acceptance angles into the spectrometer should be high enough to include a sizable fraction of the scattered electrons leaving the specimen. In the energy range of interest it should be at least 10 mrad at 100 keV incident energy.[30] For edges above 1 keV even larger acceptance angles are desirable. Spectrometer aberrations cause the energy resolution to deteriorate with increasing collection angles. Post-specimen optics can be used to match the scattering angles of the spectrometer and hence increase the angular collection efficiency while maintaining reseasonable energy resolution. Above 2 or 3 keV the collection efficiency is still however one of the main limitations on the recording of data.

Finally we should add that EXELFS can be recorded with spectrometers other than the magnetic type and in the next section we shall discuss the analysis of data from such a system. These spectra were recorded with a retarding field Wien Filter spectrometer attached to a CTEM operating at 75 keV.[31] This instrument has the flexibility of allowing us not only to measure transmitted electron intensity as a function of energy loss but also as a function of scattering angle or momentum transfer. It is therefore possible to investigate the momentum transfer dependence and anisotropy effects in EXELFS and these will be discussed later. However the range of scattering angles collected by the spectrometer is about 1 mrad so the efficiency is much less than optimum again. The brightness of the source, a heated tungsten filament is not optimized either. Despite these limitations we hope the data presented will serve to demonstrate how fine structure analysis may be carried out.

5. ANALYSIS IN PRACTICE

Edge Overlap

For the energy range 200 to 2000 eV the K and L_{23} edges of neighboring elements in the periodic table are separated by between ～50 and 200 eV. Extended energy loss fine structure from one edge may therefore be interrupted by another inner shell excitation. This problem is less important in X-ray absorption spectra since higher energies are normally used and fewer overlaps occur.

Figs. 5(a) and 5(b) show respectively the boron and nitrogen K edges in a sample of cleaved compression annealed boron nitride. Fine structure, shown both at small and large momentum transfer for each element will be discussed in detail later. For the moment we are concerned with the small amount of carbon contamination giving rise to a K edge at 285 eV superimposed on the extended fine structure originating from the boron K edge at 192 eV. Similarly a small oxygen K edge is detected at 530 eV on the tail of the nitrogen K edge at 400 eV, presumably due to impurity present in the sample. The EXELFS modulations are weak enough however at 100 eV or so above threshold for the impurity peaks to be relatively large. A complete analysis of these spectra is precluded because of the difficulty in removing the unwanted carbon and oxygen peaks. When the interfering edge is very weak in some cases it may possibly be ignored. This would be appropriate if only a small discontinuity occurred in the background-subtracted spectrum so that the result of the Fourier Transform would not be drastically affected but only contain some additional noise.

Plural Scattering

An important factor intrinsic to the electron energy loss technique is plural inelastic scattering. This becomes appreciable if the specimen thickness approaches the total inelastic mean free path or about 500Å at 100 keV incident energy. Collective

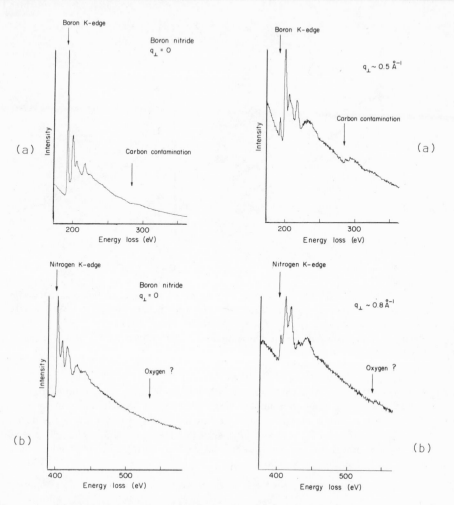

Fig. 5.   (a) Spectra from BN in the region of the boron K edge at $q_\perp = 0$ and with
            $q_\perp \sim 0.5 Å^{-1}$.
            (b) Spectra from BN in region of the nitrogen K edge at $q_\perp = 0$ and with
            $q_\perp \sim 0.8 Å^{-1}$.

plasmons or other processes involving valence electrons are the most probable excitations and these are typically peaked at energies from about 5 to 25 eV. When the incident fast electron passes through a thin sample, multiple losses occur according to a Poisson distribution in the low energy part of the spectrum. Above the core edge, the spectrum is affected too, since after exciting an inner shell the fast electron may subsequently give rise to one or more valence electron excitations. Extra peaks above threshold are produced and these may cause inaccuracies in EXELFS analysis. Plural scattering also tends to increase the background intensity relative to the signal and hence degrades the statistics.

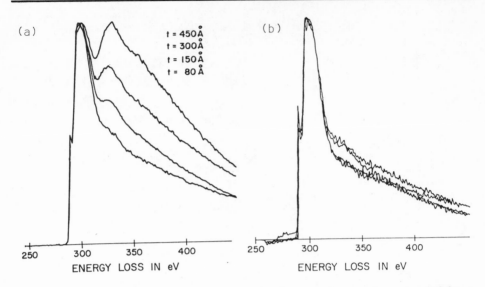

Fig. 6.   (a) Energy loss spectra at the K edge in amorphous carbon recorded from
samples of differing thickness with 75 keV incident electrons by Ray [Ref.
33].
(b) Deconvoluted spectra showing single scattering profiles.

It is important to minimize these effects either by choosing a thin enough sample or
by removing plural scattering completely. Such an approach may be considered as follows. The observed core edge spectrum $I_C(E)$ is the result of convoluting the desired
single scattering profile $I_S(E)$ with the measured low loss spectrum $I_L(E)$. The single
scattering profile may therefore be recovered by deconvoluting the observed core loss
spectrum with the measured low losses,

$$I_C(E) = I_S(E) * I_L(E) \tag{11}$$

$$\hat{I}_S(\hat{E}) = \hat{I}_C(\hat{E}) / \hat{I}_L(\hat{E})$$

where * denotes convolution and ^ denotes a Fourier Transform.

Firstly, the background intensity preceding the edge should be subtracted off using
for example a simple empirical inverse power law extrapolation.[23] It should be added
that the deconvolution method only holds rigorously if spectra are collected over all
scattering angles since the angular distributions are themselves a function of energy
loss. Fig. 6 shows the above procedure applied by Ray[33] to a series of carbon films of
differing thicknesses from 80 to 450Å. A filter is employed to reduce the effect of
noise in the deconvolution. In the experimental data a spurious peak above the carbon
K edge develops with increasing sample thickness. After deconvolution the resulting
spectra all have approximately the same shape, corresponding to the single scattering
profile. From these results it is evident that plural scattering can seriously affect the
spectrum even in quite thin samples. However for thicknesses of about 100 or 200Å at
100 keV incident energy, the spectrum should not depart significantly from the single

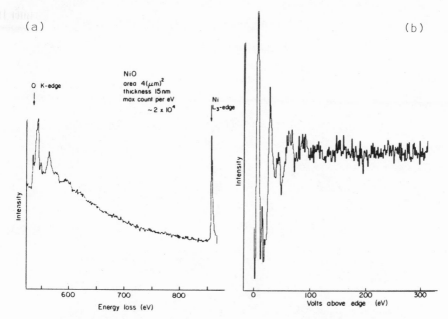

Fig. 7.   (a) Oxygen K edge from a thin NiO film, extending as far as the Ni L_3 white line.
(b) Residual EXELFS from the oxygen edge.

scattering profile and we have chosen to record data from such specimens.

It is noted that the specimen thickness can be estimated from the ratio of the zero loss intensity J_0 to the total intensity in the spectrum J_{TOT}. From the Poisson distribution we have

$$t \cong \lambda_i \, \ln \left[\frac{J_{TOT}}{J_0} \right] \qquad (12)$$

where λ_i is the total inelastic near free path. A preliminary inspection of the low loss spectrum provides a useful check on the degree to which plural scattering will affect the core edge structure. The effect of plural scattering can be reduced by increasing the incident electron energy so that the inelastic mean free path increases. It should be advantageous to work at 200 keV in most cases or perhaps at even higher energies.

K Edges

As an example of EXELFS analysis from a K edge we shall consider the spectrum of NiO in Fig. 7(a). The sample was prepared by oxidation of a thin 100Å evaporated film of the metal and spectra were recorded from an area about 4 μm in diameter, by means of a Wien Filter spectrometer combined with a transmission electron microscope.[31] Fine structure above the oxygen K edge at 530 eV extends out beyond the nickel L_{23} threshold at 855 eV, the onset of which is evident as a very strong peak attributable to a large density of 3d states. This has been referred to as a "white line" in the

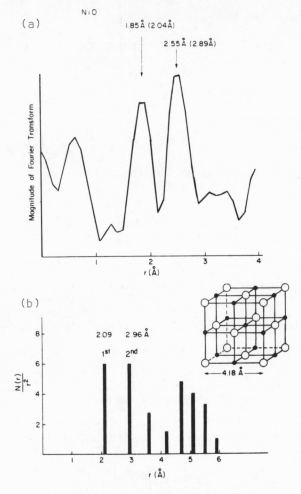

Fig. 8.   (a) Radial distribution function for *NiO*.
              (b) Expected radial distribution for *NiO* (NaCl lattice).

X-ray literature because the strong absorption gives a white line on photographic plates.[28] Although fine structure is also visible above the NiL_{23} edge, statistics are not as good as for the oxygen K edge and we prefer to analyze the oxygen edge structure rather than the nickel. After subtraction of the smooth background intensity above threshold, fitted from 600 to 850 eV with a cubic polynomial, the residual modulations in Fig. 7(b) were obtained. The intensity was multiplied by k^2 after changing the energy scale above threshold to momentum, and the data was truncated over a k-range from 2Å^{-1} to 9Å^{-1} and transformed to give the radial distribution function in Fig. 8(a). The total k-range was increased four times to carry out the Fourier Transform so that the step size in the radial distribution function was reduced. Apart from a spurious

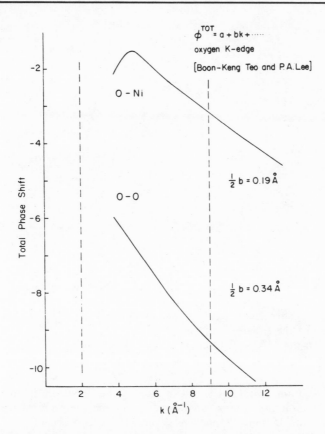

$\phi^{TOT} = a + bk + \cdots$

oxygen K-edge

[Boon-Keng Teo and P.A. Lee]

O – Ni

O – O

$\frac{1}{2} b = 0.19 \text{Å}$

$\frac{1}{2} b = 0.34 \text{Å}$

Fig. 9.   Sum of backscattered and $l = 1$ central atom phase shifts for $O-Ni$ and $O-O$ from Teo and Lee [Ref. 22].

peak at small r arising from inaccuracies in the background subtraction, the result contains two peaks of approximately equal intensity at 1.85Å and 2.55Å. The positions of these were found to be insensitive to the precise range of k chosen for the Fourier transform. Phase shifts calculated by Teo and Lee[22] for oxygen-oxygen and oxygen-nickel atom pairs are shown in Fig. 9. We have added the linear parts of these phase shifts on to the peak positions assuming the first coordination shell comprises nickel and the second shell oxygen atoms. We then obtain values of 2.04Å and 2.89Å for the nearest and second nearest neighbors; these are indicated in parenthesis in Fig. 8(a). Nickel oxide is known to have the $NaCl$ structure and this is shown in Fig. 8(b). The first two coordination shells at 2.09 and 2.96Å agree satisfactorily with the EXELFS results when the noise level and limited range of data are taken into account. The ratio of first and second coordination shell peak intensities derived from the extended fine structure is also in reasonable agreement with the known value.

L₂₃ Edges

The L_{23} edges of elements with atomic numbers below about 40 are accessible by electron energy loss spectroscopy at energies below 3 keV. We note that the K edges

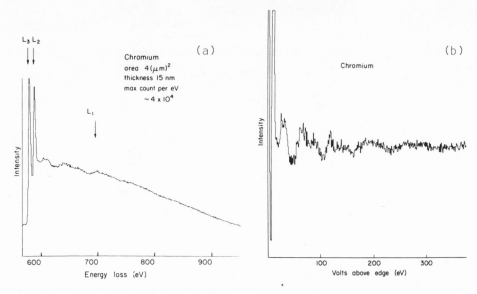

Fig. 10.   (a) L shell excitation in a chromium film showing the spin-orbit split L_{23}
edge and the L_1 edge at 696 eV.
(b) Residual EXELFS from the Cr L_{23} edge.

may be useful at atomic numbers less than 20. L_{23} fine structure therefore provides
the possibility of obtaining information about local atomic environments in medium Z
elements such as the 3d transition metals. Although the cross section for L shell excitation is favorable, the spectrum is complicated by three different excitations leaving a
$2s_{\frac{1}{2}}$, $2p_{\frac{1}{2}}$ and $2p_{\frac{3}{2}}$ hole, the L_1, L_2 and L_3 edges respectively. In X-ray absorption
spectra of the higher Z elements these edges are usually sufficiently separated for
analysis to proceed normally. For the 3d transition elements with 2p binding energies
ranging from about 400 to 900 eV, overlap effects cannot be ignored however.

Fig. 10(a) shows the L edge spectrum recorded at 75 keV incident energy from a
150Å film of evaporated chromium. L_3 and L_2 edges at 575 and 584 eV are evident as
two sharp "white lines" separated by the spin-orbit splitting of 9 eV. As a first approximation it is reasonable to neglect the L_2 edge. This can be justified on the basis that
the spin-orbit splitting is small compared with the spacing of the extended fine structure
maxima well above threshold. Also the intensity of the L_2 edge structure is expected
to be only half that of the L_3 structure because of the $(2j+1)$ degeneracy of the initial
state, where j is the total angular momentum quantum number. For the NiL_{23} edge
this approximation is less valid since the spin-orbit splitting is 17eV.

More serious interference arises from the L_1 excitation in chromium at 696 eV. As
seen in Fig. 10(a) there is no sharp edge at this energy due partly to the comparatively
large intrinsic width or short lifetime of the 2s hole.[34] Denley et al.[16] in X-ray absorption work on titanium have chosen to fit a polynomial through the broad L_1 peak.
Instead we have decided to multiply by a step function at this energy so that fits to the

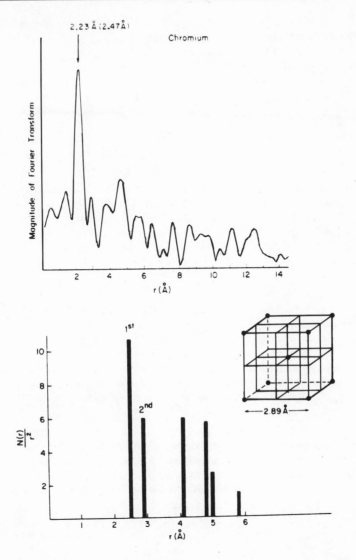

Fig. 11.   (a) Calculated radial distribution function in *Cr* showing nearest neighbor
distance.
(b) Theoretical radial distribution function for chromium.

background before and after the L_1 edge are continuous. Fig. 10(b) shows the resulting intensity modulations with this procedure applied to the chromium spectrum. After
changing to a wavenumber scale, the data is multiplied by k^2 and transformed to give
the radial distribution function in Fig. 11(a). The experimental data were truncated at
2Å^{-1} and 10Å^{-1} and transformed over a k-range four times greater than this to reduce
the step size in the radial distribution as was done for the *NiO* spectrum.

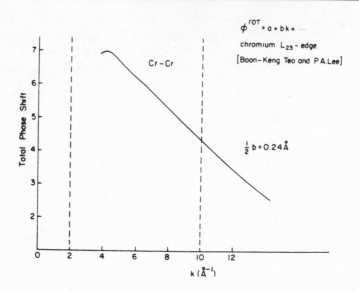

Fig. 12.   Sum of backscattered and $l = 2$ central atom phase shifts for $Cr - Cr$ from Teo and Lee [Ref. 22].

A strong peak at a distance of 2.23Å is found in Fig. 11(a) which is well above the noise level. Again we adopt the calculated phase shift of Teo and Lee[22] in Fig. 12 and take only the linear part. An upward shift in the radial distribution by this amount gives 2.47Å for the nearest neighbor distance. This agrees closely with the known radius of the first coordination shell at 2.50Å in the body-centered cubic lattice of chromium shown in Fig. 11(b).

We see that the second coordination shell of chromium is not evident in Fig. 11(a) even though the resolution should be adequate. This discrepancy may be related to our neglect of the L_{23} spin-orbit splitting. We have therefore attempted to carry the analysis one stage further by separating the L_3 and L_2 edges. To do this we suggest deconvoluting the measured spectrum, after removal of the L_1 intensity, by a pair of δ-functions spaced in energy by the spin-orbit splitting (9 eV in Cr) and suitably weighted by the theoretical L_3/L_2 intensity ratio of 2:1. This procedure assumes that the observed spectrum consists of the L_3 edge shape and, superimposed on it, a similar curve of half the intensity shifted up in energy by 9 eV. The result of the deconvolution is shown in Fig. 13(a) where the L_2 peak has almost disappeared but the L_3 first "white line" peak is over three times the intensity just above threshold. In fact the procedure does not separate the L_3 and L_2 lines completely and this seems either to reflect that their shapes are not exactly the same, or some anomaly in their ratio exists. Although some additional noise is present in Fig. 13(a) significant differences are noticeable compared with the raw data. The computed radial distribution function in Fig. 13(b) now contains a second peak at 3.0Å after the phase shift correction is applied. This may correspond to the second coordination shell in the body-centered cubic lattice at 2.9Å. Both peaks in the radial distribution are insensitive to the exact choice of k_{min} and k_{max} used to transform the data. We conclude that such an approach may be useful for analyzing L_{23} edge data.

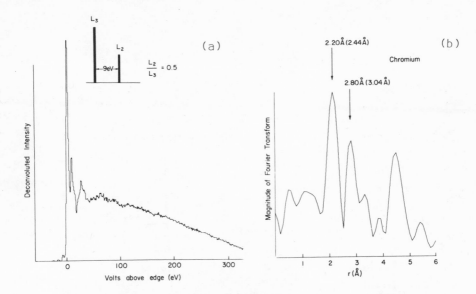

Fig. 13.   (a) Result of deconvolution of experimental L_{23} spectrum in chromium by a
pair of δ-functions spaced 9 eV apart and weighted 2:1. The L_1 intensity
was removed first.
(b) Radial distribution function obtained from deconvoluted spectrum.

6. ANISOTROPY EFFECTS

For anisotropic crystalline materials rather more detailed information concerning the
local atomic environment in different crystal directions can be extracted from extended
fine structure data, at least in principle. Anisotropic effects have been predicted
theoretically[3] and demonstrated experimentally by Heald and Stern[18] in X-ray absorption work on layer materials such as WSe_2. For K edges, anisotropy of the extended
fine structure can be considered in terms of the orientation of outgoing waves with p-
symmetry. In the case of X-ray absorption the orientation of the outgoing photoelectron depends on the electric field polarization direction $\hat{e}$ in eq. 2. If $\hat{e}$ is perpendicular
to the position vector $\underline{R}_j$ of the jth atom with respect to the central atom, the contribution of atom j will be zero, and when $\hat{e}$ is parallel to $\underline{R}_j$ the backscattered contribution
will be maximum. The orientation dependence is given from eq. 2 by,

$$\chi \, \alpha \, (\hat{\underline{e}} \cdot \underline{R}_j)^2 = \cos^2 \Phi_j \qquad (13)$$

where Φ_j is the angle between $\hat{e}$ and $\underline{R}_j$ (see Fig. 14). For L_{23} edges the orientation
dependence is more complicated since the final state can have s or d symmetry.[18]

A similar situation should hold for electron scattering since the matrix elements
have the same form, except $\hat{e}$ is now replaced by $\hat{\underline{\epsilon}}_q$ a unit vector along $\underline{q}$. (See eq. 6).
The direction of q depends or the scattering angle θ on the angle between q and the
incident beam, $\beta = \tan^{-1}(\theta/\theta_E)$, as shown in Fig. 14. In the following it is assumed
that $|\underline{q}|$ is still small enough for the dipole approximate to be valid. The orientation

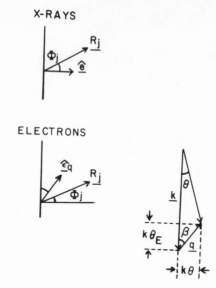

X-RAYS

ELECTRONS

Fig. 14.   Diagram showing orientation of the atom at $\underline{R}_j$ with respect to $\hat{e}$ and $\hat{\varepsilon}_q$ for X-ray absorption and electron scattering respectively. The relation between $\hat{\varepsilon}_q$ and the scattering angle θ is also indicated.

dependence of the EXELFS for K-edges is then given by,

$$\chi \ \alpha \ \sin^2(\Phi_j + \beta) \tag{14}$$

In order to give an example of anisotropy effects we return to the spectra in Figs. 5(a) and 5(b) from the highly anisotropic layer structure, boron nitride which is similar to graphite. These were recorded with the c-axis oriented parallel to the incident beam. Large changes in near edge structure can be explained in terms of transitions to unoccupied bands with π or σ character above the Fermi Level.[35] Sharp peaks can be attributed to high densities-of-states, modified by excitonic effects. When the momentum transfer suffered by the incident electrons is parallel to the π^* antibonding states directed along the c-axis a strong $Is \rightarrow \pi^*$ peak is observed. This occurs at zero scattering angle (Fig. 14) when the sample is oriented with c-axis parallel to the incident beam. For larger scattering angles q is almost perpendicular to the c-axis and the $1s \rightarrow \sigma^*$ transitions are evident.

As discussed above, an anisotropy dependence is also expected for the extended fine structure. Fig. 15 shows the EXELFS above the boron K edge at two values of perpendicular momentum transfer, $q_\perp \sim 0$ and $q_\perp \sim 0.5\text{Å}^{-1}$. The background intensity has already been removed and the data extend as far as the carbon K edge at 285 eV. At $q_\perp \sim 0(\theta = 0)$ the momentum transfer is along the incident beam direction and the c-axis of the boron nitride. Atoms in the layer planes should therefore contribute little to the intensity modulations. For large θ, q is nearly contained in the layer planes and these atoms should contribute strongly. An observed decrease in the EXELFS amplitude at zero scattering angle in Fig. 15 compared with the amplitude at larger q may be

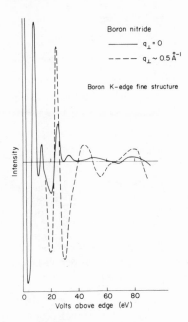

Fig. 15.   Residual EXELFS intensity for the boron K-edge in hexagonal boron nitride
at $q_\perp = 0$ and $q_\perp = 0.5\mathring{A}$ from Fig. 5(a).

explained by the greater distance between layer planes, over twice the nearest intra-
planar spacing.

Although the occurrence of the carbon K edge in Fig. 5(a) precludes a full analysis
of the data because of the limited energy range available, Fig. 15 is consistent with an
increased contribution from atoms at about 3Å distance in the fine structure recorded
with q along the c-axis. This agrees with the known structure (nearest neighbor distance of 1.4Å and interplanar spacing of 3.3Å).

7. SUMMARY

We have demonstrated that inelastic electron scattering offers a useful alternative to
X-ray absorption spectroscopy for measuring core edge fine structure is some materials
and under certain conditions.

The main advantages to have emerged are:

1. EXELFS is sensitive to the lower atomic number elements, a range $5 \lesssim Z \lesssim 20$
   for the K edges and $20 \lesssim Z \lesssim 40$ for the L_{23} edges.

2. The technique can be used with very small electron probes so that data may be collected from microscopic regions in a thin sample which may be prepared as for
   electron microscopy.

3. EXELFS can be combined with other techniques such as electron diffraction or
   high resolution imaging in the " analytical electron microscope." It is important to

realize that often no single technique can provide all the answers but that a more complete understanding may be attained using several techniques together.

Some of the problems associated with the EELS technique are:

1. EXELFS data are collected at lower energies than EXAFS so that overlapping edges are more likely to complicate analysis.

2. Samples must be very thin, a few hundred Å otherwise plural scattering effects become difficult to remove.

3. Estimated statistics suggest that an atomic species needs to be present in reasonably high concentration, probably greater than about 20%.

The similarity of the cross sections for inelastic scattering and X-ray absorption makes the same type of analysis applicable to each technique. We have shown too how the dependence of the energy loss spectrum on momentum transfer might be exploited to give information about crystal anisotropy.

The application of the electron technique to some classes of problems in materials science seems to be promising, particularly to disordered materials containing light elements. Finally data which we have described can no doubt be improved considerably since little effort has yet been made to optimize the experiments.

ACKNOWLEDGEMENTS

The authors would like to thank Dr. P. Batson and Professor M. Isaacson for useful discussions. Financial support from the National Science Foundation, grant number DMR-78-09204, through the Materials Science Center at Cornell is gratefully acknowledged.

REFERENCES

1. F. W. Lytle, in Advances in X-ray Analysis, edited by G. R. Mallett, M. Fay and W. M. Mueller (Plenum, New York) 9, p. 398 (1966).

2. D. E. Sayers, E. A. Stern and F. W. Lytle Phys. Rev. Lett., 27: 1024 (1971).

3. E. A. Stern, Phys. Rev. B10: 3027 (1974).

4. F. W. Lytle, D. E. Sayers and E. A. Stern, Phys. Rev. B11: 4825 (1975).

5. E. A. Stern, D. E. Sayers and F. W. Lytle, Phys. Rev. B11:: 4836 (1975).

6. J. J. Ritsko, S. E. Schnatterly and P. C. Gibbons, Phys. Rev. Lett., 32: 671 (1974).

7. R. D. Leapman and V. E. Cosslett, J. Phys. D9: L29 (1976).

8. C. Colliex, V. E. Cosslett, R. D. Leapman and P. Trebbia, Ultramicroscopy, 1: 301 (1976).

9. B. M. Kincaid, A. E. Meixner and P. M. Platzman, Phys. Rev. Lett., 40: 1296 (1978).

10. P. E. Batson and A. J. Craven, Phys. Rev. Lett., 42: 893 (1979).

11. M. Isaacson and M. Utlaut, Optik, 50: 213 (1978).

12. U. Fano and J. W. Cooper, Rev. Mod. Phys. **40**: 441 (1968).

13. H. A. Bethe, Ann. Phys. (Leipzig) **5**: 325 (1930).

14. M. Inokuti, Rev. Mod. Phys. **43**: 297 (1971).

15. J. Stöhr, R. S. Williams, G. Apai, P. S. Wehner and D. A. Shirley, Proceedings of the Fifth International Conference on VUV Radiation Physics, Montpellier, September 1977, p. 43.

16. D. Denley, R. S. Williams, P. Perfetti, D. A. Shirley and J. Stöhr, Phys. Rev. **B19**: 1762 (1979).

17. J. H. Hubbell, At. Data **3**: 241 (1971).

18. S. M. Heald and E. A. Stern, Phys. Rev. **B16**: 5549 (1977).

19. R. F. Egerton, Ultramicroscopy **3**: 243 (1978).

20. P. A. Lee and J. B. Pendry, Phys. Rev. **B11**: 2795 (1975).

21. P. A. Lee and G. Beni, Phys Rev. **B15**: 2862 (1977).

22. B. Teo and P. A. Lee, J. Am. Chem. Soc., **101**: 2815 (1979).

23. R. F. Egerton, Phil. Mag., **31**: 199 (1975).

24. B. M. Kincaid and P. Eisenberger, Phys. Rev. Lett. **34**: 1361 (1975).

25. C. J. Rossouw, D. Phil. Thesis, University of Oxford, England (1977).

26. P. E. Batson, J. M. Gibbons and R. P. Ferrier, Proceedings of the 8th International Conference on Amorphous and Liquid Semiconductors, Boston, August 1979 (in press).

27. D. E. Johnson and S. Csillag, Proceedings of the 37th EMSA Meeting, San Antonio, August 1979, ed G. W. Bailey, (Claitors Press: Baton Rouge) p. 526.

28. R. D. Leapman, P. Rez and D. F. Mayers, J. Chem. Phys. **72**: 1232 (1980).

29. D. E. Johnson, Ultramicroscopy **3**: 361 (1979).

30. D. C. Joy and D. M. Maher, Ultramicroscopy **3**: 69 (1978).

31. G. H. Curtis and J. Silcox, Rev. Sci. Instrum. **42**: 630 (1971).

32. R. F. Egerton and M. J. Whelan, Phil. Mag. **30**: 739 (1974).

33. A. B. Ray, Proceedings of the 37*th* EMSA meeting, San Antonio, August 1979, ed G. W. Bailey, (Claitors Press: Baton Rouge) p. 522.

34. O. Keski-Rahkonen and M. O. Krause, At. Data **14**: 139 (1974).

35. R. D. Leapman and J. Silcox, Phys. Rev. Lett., **42**: 1361 (1979).

EXTENDED ENERGY LOSS FINE STRUCTURE STUDIES

IN AN ELECTRON MICROSCOPE

S. Csillag[*], D. E. Johnson[**], E. A. Stern[***]

*Department of Physics
University of Stockholm
Stockholm, Sweden   11346

**Center for Bioengineering, RF-52
University of Washington
Seattle, Washington   98195

***Department of Physics, FM-15
University of Washington
Seattle, Washington   98195

INTRODUCTION

Extended fine structure similar to extended X-ray absorbtion fine structure has been observed in electron energy loss experiments.[1-4] The modulations in the differential inelastic electron scattering cross-section past an absorbtion edge are called Extended Energy Loss Fine Structure (EXELFS) and contain information equivalent to EXAFS.

In this chapter we present a more extended feasibility study of this technique in the electron microscope pointing out some of the similarities and the differences between EXAFS and EXELFS. The EXELFS data are obtained from spatially resolved areas which can be characterized using electron micrographs and electron diffraction patterns.

EXAFS is the modulation in the photo-absorbtion cross-section due to the interference between the outgoing excited inner shell electron waves and the electron waves backscattered from the surrounding atoms. The inner shell electrons in EXAFS are excited using synchroton X-rays. However, at the lower energies the X-rays have very little penetrating power. This causes major problems in studying low Z-materials with EXAFS.

Another major limitation which has to be taken into account is the inconvenience of traveling to a synchroton source. These limitations have encouraged a search for alternative methods to measure EXAFS. From earlier experiments done with the electron energy loss technique, it is known that the distribution of energy lost by a beam of electrons passing through a thin foil of material is a direct measure of the electronic excitations produced by the beam in the sample. A monoenergetic incoming electron beam has three classes of regions in the energy distribution after the passage through the sample.

a. The unscattered electrons which pass through the sample without losing energy. This is the so-called primary beam or zero loss peak.

b. The inelastically scattered electrons which produce valence shell excitations such as surface or bulk plasmons and lose the corresponding amount of energy necessary to produce this type of excitation. The energy range for these types of excitations is typically between 7-30 eV.

c. The inelastically scattered electrons which produce inner shell excitations losing energies necessary to produce K, L and M shell ionizations.

The number of electrons in each of the these regions are proportional to the probability that each of these energy loss events occurs. The probability for these events is a function of the incoming electron energy, the thickness, and the composition of the sample. Figure 1 shows a typical energy loss spectrum obtained using a thin film of graphite (estimated thickness c.a. 300Å). During such an inelastic scattering event, the bombarding electrons excite inner shell electrons, with the same interference phenomenon between the outgoing and backscattered part of the excited electron as for the photoelectron in the EXAFS case. This means that the probability for a certain energy loss is increased or decreased depending on whether constructive or destructive interference occurs. This produces modulations in the differential inelastic electron scattering cross-section which are very similar to the modulations in the photoabsorbtion cross-section in the EXAFS case.

BASIC FORMULISM

The probability of X-ray absorption by an inner shell electron is proportional in the dipole approximation by:

$$| <f \mid \vec{r} \cdot \vec{\epsilon} \mid i > |^2 \, \rho(E_f) \qquad (1)$$

where $i>$ is the inner shell initial state, $|f>$ is the final unoccupied state. $\vec{r}$ is the position operator, $\vec{\epsilon}$ is the electric field vector, $\rho(E_f)$ is the final state density.

The energy ΔE, to low Z atoms lost by the inelastically scattered electrons is much smaller than the total energy of highly energetic bombarding electrons. This implies that the Born approximation can be used, and the matrix element in the transition rate is:

$$| <f \mid e^{i\vec{q} \cdot \vec{r}} \mid i > |^2 \rho(E_f) \qquad (2)$$

where $\hbar \vec{q}$ is the momentum transfer, $|i>$ is the initial state $|f>$ is the excited final state of the atom. The difference in energy between these states $= \Delta E$, the energy loss of the bombarding electrons.

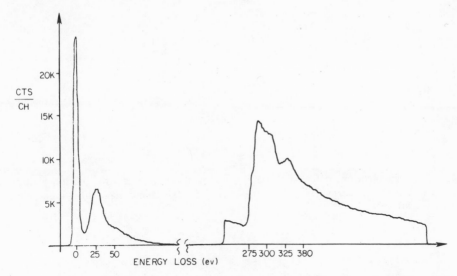

Fig. 1.   Low lying and carbon K edge spectra from a thin film of graphite. The
modulations past the 285 eV edge are due to the EXELFS. Incident energy
= 100 keV, accumulation time = 1 sec/data point ($\sim$250 sec), beam
current = $10^{-9}A$, ($\sim 10^{-12}A$ — low lying spectra) acceptance angle = 5mr.

The momentum transfer q for particles of incident momentum q_i is determined by
the scattering angle, $q = 2q_i \sin \frac{1}{2} \theta$, as it is shown in figure 2.

If $|i>$ is a core state and if $q \approx 0$ (forward or nearly forward scattered electrons)
then $qr_c \ll 1$ where r_c is the radius of the initial core state, and if the electrons are
independent, expression (2) can be written as:

$$q^2 < f|\hat{\epsilon}_q \cdot \overline{r}|i > |^2\rho(E_f) \tag{3}$$

where q is the momentum transfer and $\hat{\epsilon}_q$ is the unit vector in the q direction. This
means that for small $\overline{q}$ values the dipole approximation is valid similar to X-ray absorption.

Thus for small q, EXAFS and EXELFS contain the same information.

Quantum mechanically both cases show the same variation in the matrix element
with energy leading to the fine structure in EXAFS and in EXELFS. Because the electron scattering cross section in the dipole approximation is proportional to the photoabsorption cross-section, we can look at the EXELFS as the electron analog of the
EXAFS. There is an additional q dependence in expression 2 which can be used to
obtain additional near edge structure information and information about the final state
symmetry in crystalline materials.

The EXAFS formula given below for the modulation in energy absorbtion as a function of the ejected photoelectrons wave vector is also valid for EXELFS in the dipole
approximation.

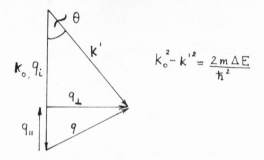

Fig. 2.   An electron scattered through an angle θ by an inelastic collision giving a loss of energy $\Delta E (= \hbar \omega)$ will also undergo a corresponding momentum transfer $\hbar q$.

$$k\chi(k) = A \sum_j \frac{N_j}{R_j^2} t_j (2k) \sin 2[kR_j + \delta_j(k)]\exp(-2R_j/\lambda) \exp(-2k^2\sigma_j^2) \quad (4)$$

With A = a constant and $\chi(k) = (I - I_o)/I_o$ where I = Intensity of Modulations and I_o = Absorbtion Edge Intensity

The first exponential decay term is due to inelastic scattering of the ejected electrons (λ = mean free path) and the second is due to deviations of the atomic positions (σ = root mean deviation) about their average position, both of which tend to decrease the amplitude of the modulations.

A is a constant and the sum is over the various j^{th} coordination shells containing N_j atoms at the average distance R_j from the absorbing atom, $t_j(2k)$ is the amplitude for backscattering from the jth atom, and $\delta_j(k)$ is the phase shift. This phase shift is a sum of terms from both the central and the scattering atoms and an effective phase shift term resulting from error in determining the threshold energy (6).

ADVANTAGES OF EXELFS

The major advantages using high energy electrons instead of synchrotron X-rays can be summarized as follows (For a detailed discussion see ref. 7):

a. EXELFS permits the extraction of fine structure information from low Z elements, which is a difficult problem for synchrotron radiation EXAFS.

b. One can focus the electron beam to very small areas (spot sizes of about 100Å $\times$ 100Å) which gives spatial resolution allowing the opportunity to study inhomogeneous samples.

c. Using the electron microscope one can image the irradiated area and also obtain the diffraction pattern of the sample both of which contain useful additional information.

d. The data gathering time is comparable to that of synchrotron sources.

e. One can study the momentum transfer dependence of the inelastic electron scattering cross-section.

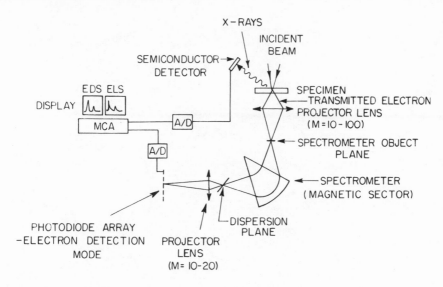

Fig. 3.   A schematic diagram of the instrumentation used for data collection and initial processing.

f. Finally, the instrumentation is more accessible and less expensive than synchrotron sources.

These substantial advantages compared to EXAFS makes a more extended feasibility study of this technique extremely worthwhile.

DATA GATHERING

The experimental apparatus used in our measurements consists of a JEOL 100C transmission electron microscope with scanning attachment and magnetic sector electron energy loss spectrometer, interfaced to a Kevex 7000 computer based multichannel analyzer for data gathering and initial processing.

The design and construction of the energy loss spectrometer and its interfacing to the JEOL 100C CTEM/STEM and Kevex 7000 are described in detail elsewhere (8) and only a brief summary of the characteristics are given here.

The spectrometer is a straight edge magnetic (double focusing) spectrometer with a radius of curvature = 5 cm, uniform magnetic field = 224 gauss at 100 keV, dispersion = $1 \mu m/eV$ at 100 keV, bending angle = 90°, object and image distances = 10 cm, magnification = 1. Electrons of a given energy loss are selected by a variable width slit and counted using a scintillator-photomultiplier-discriminator combination which sends pulses to a multi-channel analyzer, operated in the time sequence store mode. The spectrometer current is ramped synchronously with the sweep through the analyzer channels thus producing the energy loss spectrum. A schematic diagram of the instrument is shown in figure 3.

In order to demonstrate the feasibility of the technique, we have studied three different systems with four different ionization edges, namely graphite, Al, and Al_2O_3

Fig. 4a.   A transmission electron micrograph of a small area of an evaporated *Al* film
            used in this work. The area is ~3000Å by ~3000Å and represents the
            approximate size of the regions analyzed. The specimen is clearly polycry-
            stalline with diffraction contrast evident, consistent with the polycrystalline
            electron diffraction ring patterns obtained from the same specimens.

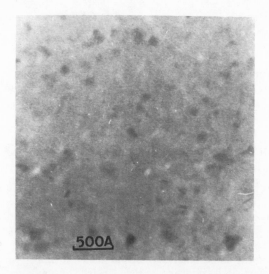

Fig. 4b.   A transmission electron micrograph of a small area of an Al_2O_3 film used in
            this work. The area is ~3000Å by ~3000Å and represents the approximate
            size of the regions analyzed. The specimen shows only small mass thickness
            variations and no diffraction contrast consistent with a norphous electron
            diffraction patterns obtained from the same specimens.

(both Al and O, K edges). The thin films required in this technique have been prepared in different ways. The graphite film was prepared by cleaving thin graphite layers with adhesive tape (estimated thickness approx. 300Å).

The thin Al film (thickness 200Å) was evaporated onto a Formvar substrate. The Al_2O_3 sample was prepared as a thin Al layer evaporated on the surface of a freshly cleaved $NaCl$ crystal at 20°C. The film was oxidized by heating in air. Al_2O_3, being an insulator, charges up during exposure to the electron beam and in order to avoid this, we evaporated a thin (~30Å) carbon film onto the Al_2O_3. Floating the film off the $NaCl$ in water, we obtained a thin, carbon-coated Al_2O_3 film with an estimated thickness of about 200Å. Figures 4a and 4b are transmission electron micrographs of the thin Al and Al_2O_3 film respectively. The areas imaged represent the approximate size (~3000Å × 3000Å) of the areas from which the EXELFS data reported here were gathered. The micrographs demonstrate the polycrystalline nature of the Al films with clear diffraction contrast, while the Al_2O_3 films show only mass thickness contrast indicative of an amorphous nature. Electron diffraction patterns obtained from these same specimens showed polycrystalline ring patterns from the Al films and diffuse amorphous patterns from the Al_2O_3 films. The electron micrographs, electron diffraction patterns and the electron energy loss spectra reported here were all obtained in the same instrument and on the same part of the sample. Figures 5a, and 5b show typical EXELFS past the Al K edge and past Al and O K edges in Al_2O_3. The data were obtained with an accumulation time of 1 sec/data point for a total time of approx. 250 sec. The incident beam energy = 100 keV, the beam current was $4 \times 10^{-9}A$ and the acceptance half angle = 9 mr (0) and 18 mr (Al).

The instrumental energy resolution for the above spectra was approximately 8 eV determined mainly by lens aberrations and the size of the irradiated area (9). With a more efficient detection system, the beam current and thus the size of the irradiated area could be reduced resulting in an energy resolution of approximately 2-3 eV. By using alternative electron sources (i.e. LaB_6 or field emission Tungsten) the resolution could be further reduced to 0.5 to 1.0 eV (10).

DATA ANALYSIS AND RESULTS

The data analysis involved several steps: 1) smooth background subtraction, 2) normalization of the modulations to the edge step height and 3) converting energy loss to ejected electron wave vector. The useful data extraction should not start closer than 30 eV past the edge, measured from the inflection point of the edge. There is fine structure information in the near edge structure also, but the EXELFS here is masked by band structure effects. For the Al film data, the analysis was started at 40 eV and for the graphite 50 eV past the edge to reduce any effect of plural scattering.

Figure 6a and 6b show the typical $k\chi(k)$ data from the carbon, and oxygen K edges of figures 1 and 5. The zero of the ejected electrons wave vector was taken at the inflection point of the edge.

After Fourier transforming the $k\chi(k)$ data, the magnitude of the transforms gives peaks centered around the distances which correspond to the different spacings in the sample. Determining the distances between the origin (where the absorbing atom is located) and the main peaks, and making the phase shift corrections, one obtains the interatomic distances, or in this case, knowing the interatomic distances accurately, one can calculate the phase shifts.

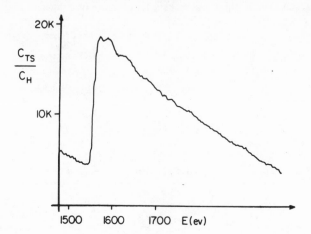

Fig. 5a.   Typical EXELFS past the *Al K* edge in the energy loss spectrum of a thin
            Al film. The incident energy was 100 keV, the half angle of acceptance was
            18 mr and the data gathering time was 200 seconds.

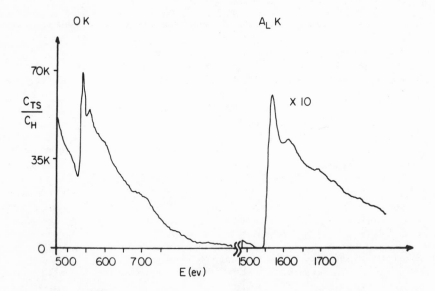

Fig. 5b.   Typical EXELFS past the *O K* and the *Al K* edges in the energy loss spec-
            trum of a thin Al_2O_3 film. The incident energy was 100 keV, the half angle
            of acceptance was 9 mr for the *O K* edge adn 18 mr for the *Al K* edge, and
            the data gathering time for each spectrum was 200 seconds.

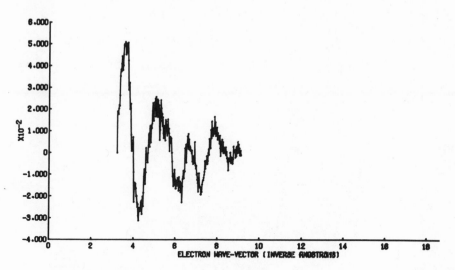

Fig. 6a. The $k\chi(k)$ data for the carbon K edge in the energy loss spectrum of a thin graphite film.

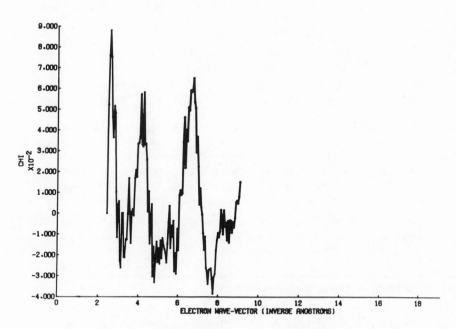

Fig. 6b. Typical $k\chi(k)$ data for the O K edge in the energy loss spectrum of a thin Al_2O_3 film. The experimental conditions are as in Figure 5.

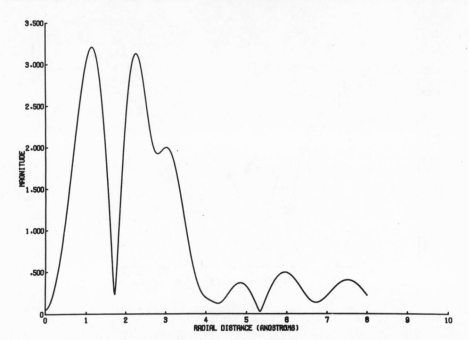

Fig. 7a.   The magnitude of the Fourier transform of the graphite $k\chi(k)$ data. The transforms are not corrected for the phase shift. The k interval was from $k = 3.22\text{Å}^{-1}$ to $k = 9.13\text{Å}^{-1}$.

The magnitudes of the Fourier transforms of the EXELFS data shown in figures 6 are plotted in figures 7. The deviations between the spacings as determined from the peaks in the Fourier transform and the true spacing should be due to a combination of k dependent phase shifts of the central and scattering atoms and an effective phase shift (also k dependent) due to any error ($\equiv \Delta E_0$) in the choice of the zero of the ejected electrons wave vector (4). A number of different techniques can be used to incorporate these terms into the extended fine structure analysis (6). For the purposes of this feasibility study, these detailed procedures were not considered necessary. A more detailed analysis of the obtained results is to be given elsewhere (11).

The main conclusion to be drawn from the results presented here is that EXELFS data gathered in relatively short times from spatially resolved areas in the electron microscope, is of sufficient quality to allow accurate EXELFS analysis.

DIFFERENCES BETWEEN EXAFS AND EXELFS

There are also significant differences between EXELFS and EXAFS. The energy of the EXAFS spectra is determined directly by the incident photon energy, while in the EXELFS case the energy loss is not varied by the incident electron energy but varies in intensity in a rapidly decreasing manner with increasing energy loss. Also, basically only the electrons for which the momentum transfer is zero or very small, carry exactly the fine structure EXAFS information of X-ray absorption. For high momentum

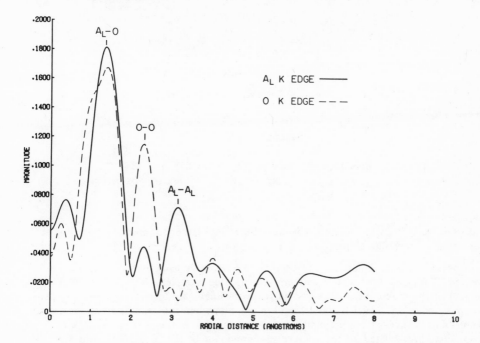

Fig. 7b.   Dashed Curve:
The magnitude of the Fourier transform of $k\chi(k)$ data from the $Al\ K$ edge in the energy loss spectrum Al_2O_3 of Figure 4. The k interval was from $k = 2.43\mathring{A}^{-1}$ to $k = 8.85\mathring{A}^{-1}$.

Solid Curve:
The magnitude of the Fourier transform of $k\chi(k)$ data from the $Al\ K$ edge in the energy loss spectrum of Al_2O_3 of Figure 4. The k interval was from $k = 3.17\mathring{A}^{-1}$ to $k = 7.77\mathring{A}^{-1}$.

transfer, the dipole approximation is no longer valid and the higher multiple contributions contribute to modify the energy loss spectrum.

For the range of scattering angles collected, the range of momentum transfers (9) was calculated and used to estimate the magnitude of the expansion parameter $(q \cdot r)$ which should be $\ll 1$ for the dipole approximation to be valid. The magnitude of r was estimated as a/Z (with $a_0 =$ Bohr radius and $Z =$ atomic number). These values of $q \cdot r$ are given in Table 1 along with the corresponding values for photon absorbtion (in which $K =$ photon wave vector magnitude). The smallest values of $q \cdot r$ for electrons corresponds to the minimum momentum transfer at zero scattering angle, and the largest value, corresponds to momentum transfer at the largest scattering angle collected.

The average $q \cdot r$ was calculated for the range of scattering angles collected by using the angular scattering distribution $d\sigma/d\Omega = (\theta^2 + \theta_E^2)^{-1}$ with $\theta_E = [(1-\beta^2)/\beta^2] \cdot E/mc^2$ ($\theta =$ scattering angle, $\beta = v/c$, $E =$ energy loss) valid near the

TABLE I

	$q \cdot r$	$k \cdot r$
	Electrons	*Photons*
Oxygen	.035 to .135 Average = .066	.019
Aluminum	.063 to .150 Average = .10	.035

absorption edge. Even though the values of $q \cdot r$ are larger for electron scattering than for photon absorbtion, they are sufficiently small to allow neglect of higher order terms.

The additional q dependence in expression (2) can be used to obtain valuable information about the final state symmetry in oriented samples. The longitudinal component of the momentum transfer contributes to final state transitions which are parallel to the beam axis, such as the π band in graphite when the beam is along the C-axis. Transitions to final states perpendicular to the beam axis (such as σ bands with P_{x-y} symmetry) are isolated by the transverse component of the momentum transfer. For small scattering angles the momentum transfer is essentially along the beam axis so that for small values of the energy loss $q_{min} \simeq q \simeq m\Delta E / \hbar q_i$. For increased values of the scattering angle the probability for σ transitions is increased which results in a less sharp and broadened main ionization peak in the case of the graphite samples.

The multiple scattering effects which occur using EXELFS are another important distinction. The probability P for only a single inelastic-scattering event occurring, decreases exponentially with the sample thickness:

$$P \propto e^{-t/\lambda}$$

t = specimen thickness

λ = mean free path for all other scattering events

We found very clear evidence of multiple scattering effects especially in graphite resulting in convolution of the EXELFS with the multiple scattering peaks. In order to avoid multiple scattering effects, one has to either use noise insensitive deconvolution programs[12] or very thin films for the measurements. In this analysis, we minimized these effects by using thin films and by beginning the analysis sufficiently past the edge to avoid the main multiple scattering peak.

SUMMARY OF LIMITATIONS OF THE EXELFS TECHNIQUE

a. In the X-ray EXAFS the sample thickness affects only the amplitude of the modulations (typically 3-5% of the edge step height) but in the EXELFS case the thickness affects both the amplitude and can add additional peaks to the modulations. This is why extra care has to be taken concerning the sample thickness. There is an optimal sample thickness depending on the energy of the bombarding electrons.

b. The maximum specimen dose (electrons/cm^2) may be limited for sensitive specimens by radiation damage.

c. The maximum specimen dose may be limited by specimen contamination if other than ultra-high vacuum systems are used.

d. The maximum data range may be limited by K ionization edges which are too close to each other such as carbon and oxygen K edges.

e. If the energy loss measurements are made at high angular resolution, for example in the study of anisotropic specimens, signal levels will decrease significantly over those found in this work using large angles of collection.

f. The probability of a given energy loss, ΔE, drops rapidly with the magnitude of ΔE, and significant EXELFS signals are limited to low Z materials, typically to the second and third row in the periodic table.

Limitation (a) above can be eliminated by the use of sufficiently thin specimens as have been used in this work. The effects of limitations b, c, and e above can all be reduced by the development and application of parallel data collection systems (i.e. all energy loss intervals detected simultaneously). Such a system could have increased the counting rates in the work reported here by a factor of $\simeq 200$. Such an increase in sensitivity can then be used to reduce the specimen dose, to increase the angular resolution, or to study time dependent processes with EXELFS. The detection of Bragg reflected electrons can open the possibility of performing surface studies from spatially resolved areas with EXELFS. Considering the substantial advantages of EXELFS compared to EXAFS and from the promising results obtained using EXELFS, it would appear that the EXELFS technique can be a very important tool in studying atomic structure in inhomogeneous complexes with low Z elements such as biological samples. In order to accomplish the same spectacular results with EXELFS as achieved in EXAFS, we may have a long way to go, but, "All grown ups were small once, although very few remember it." (The Little Prince).

ACKNOWLEDGEMENTS

This work was supported by N. I. H. grants: HL21371 and HL00472 (RCDA)-[D. Johnson and S. Csillag], by a Swedish Government scholarship (S. Csillag) and PCM 79-03674 (E. A. Stern).

REFERENCES

1. Kincaid, B. M., Phys. Rev. Lett. **40,** 19 (1978).

2. Batson, D. E. and A. J. Craven, Phys. Rev. Lett. **42,** 893 (1979).

3. Leapman, R. D. and v. E. Cosslett, J. Phys. D. Appl. Phys., **9,** (1976).

4. Johnson, D. E., Csillag, S., Stern, E. A., Proc. 37th Ann. Meeting EMSA (Claitos Press: Baton Rouge) 526 (1979).

5. Stern, E. A., Phys. Rev. B **10,** 3027 (1976).

6. Lee, P. A. and G. Beni, Phys. Rev. B **15,** 2862 (1977).

7. Isaacson, M. and M. Utlaut, Optic. **50,** 213 (1978).

8. Johnson, D. E., Rev. Sci. Inst. **51,** 705 (1980).

9. Johnson, D. E., Ultramicroscopy **5,** 163 (1980).

10. Silcox, J. (Private Communication).

11. Csillag, S., Johnson, D. E., Stern, E. A. (Unpublished).

12. Ray, A. B., Proc. 37th Ann. Meeting EMSA (Claitos Press: Baton Rouge) 522 (1979).

A COMPARISON OF ELECTRON AND PHOTON BEAMS FOR OBTAINING INNER SHELL SPECTRA[1]

M. Utlaut

The Enrico Fermi Institute and Department of Physics
The University of Chicago

I. INTRODUCTION

Recently there has been considerable interest shown in the use of synchrotron radiation as a tunable X-ray source[2,3] and electron energy loss spectroscopy within the electron microscope[4,5] for performing chemical and electronic spectroscopy of matter. It is worthwhile to compare these two kinds of radiation as to their sensitivity and applicability for studying different systems.

We will compare the following quantities: 1) the relative cross sections for the processes of interest (i.e. the photoabsorption cross section and the differential electron energy loss cross section); 2) the specific brightness of the sources since this determines the number of incident particles/sec/unit area/sterad/unit energy bandwidth which can impinge upon the specimen; 3) size of the beam (this is important if there is interest in local inhomogeneities and not in spatially averaged inhomogeneities); and 4) the effect of beam induced damage on the spectroscopic determination.

For simplification in this comparison, we will assume that interest lies mainly in probing inner shell excitations and ionizations and that the energy levels we are probing are not much deeper than several keV below the vacuum level.

II. RELATIVE SCATTERING CROSS SECTIONS

The absorption coefficient for photons of energy E within the context of dielectric theory[6] is given as

$$\mu(E) = \frac{2E}{\hbar c} \sqrt{\frac{1}{2}(|\epsilon| - \epsilon_1)} \tag{1}$$

where the complex dielectric constant $\epsilon = \epsilon_1 + i\epsilon_2$ (which is dependent on energy and momentum transfer) is taken for zero momentum transfer. If we consider only inner shell excitations where $\epsilon_1 \approx 1$ and $\epsilon_2 \ll 1$, then the photoabsorption cross section can be written

$$\sigma_\gamma(E) = \frac{1}{n}\frac{E}{\hbar c}\epsilon_2 \tag{2}$$

where n is the number of atoms or molecules per unit volume.

The differential cross section per unit energy for an electron of energy E_0 to lose an energy E in traversing an object of atomic or molecular density n is given non-relativistically as[7,8]

$$\frac{d\sigma(E)}{dE} \cong \frac{1}{2\pi a_0 E_0 n}\left[-Im\frac{1}{\epsilon(E)}\right]\ln\left(\frac{4E_0}{E}\right) \tag{3}$$

where a_0 is the Bohr radius and $\epsilon(E)$ is taken at zero momentum transfer. Equations (1) and (3) are general and measurements of $d\sigma/dE$ can be used to determine $\mu(E)$ from the Kramers-Kronig relations.[4] In the case where $\epsilon_1 \approx 1$ and $\epsilon_2 \ll 1$ since $\epsilon_2 \approx -Im\left(\frac{1}{\epsilon}\right)$, equation (3) reduces to[9,10]

$$\frac{d\sigma(E)}{dE} \approx \frac{\epsilon_2}{2\pi a_0 E_0 n}\ln\left(\frac{4E_0}{\overline{E}}\right) \tag{4}$$

and we see that $\sigma_\gamma(E)$ and $d\sigma(E)/dE$ are proportional to ϵ_2.

The ratio of photoabsorption cross section to the electron energy loss cross section for a given energy loss E is

$$\frac{\sigma_\gamma(E)}{\dfrac{d\sigma(E)}{dE}} = \frac{2\pi a_0}{\hbar c}\frac{E_0 E}{\ln\left(\dfrac{4E_0}{E}\right)} \tag{5}$$

which reduces to

$$\frac{\sigma_\gamma}{\dfrac{d\sigma}{dE}} = \frac{1.68\times10^{-3}EE_0}{\ln\left(\dfrac{4E_0}{E}\right)} \quad (E, E_0 \text{ in } ev) \tag{6}$$

If we make our comparison for 100 keV electrons, we then get

$$\frac{\sigma_\gamma}{d\sigma/dE} = \frac{170E}{\ln\left(\dfrac{4\times10^5}{E}\right)} \tag{7}$$

The photoabsorption cross section is several orders of magnitude greater than the electron energy loss cross section (fig. 1), so that for electron beam spectroscopy (at zero momentum transfer) to be competitive with photoabsorption spectroscopy, more incident electrons than photons are required to record a spectrum with a given signal to noise ratio.

III. SENSITIVITY

a. Detectability

One method of comparison of sensitivities of the two techniques is to determine the number N of atoms or molecules of a given type which can be detected (in the absence of background) with a counting rate R. This can be written

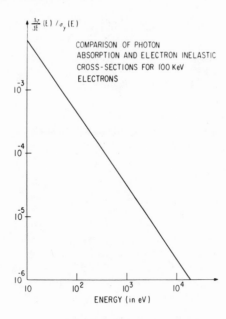

Fig. 1.   The ratio (Eq. 7) of the differential electron energy loss cross section with respect to energy to the photoabsorption cross-section. (The horizontal scale is the energy of the photons or the energy lost by the incident electrons).

$$N = \frac{R}{J\,\sigma y} \qquad (8)$$

where J is the incident particle density per unit time in the probe, σ is the cross section for the particular process to be detected, and y is the detection efficiency. Thus for a minimum practical counting rate, the sensitivity depends upon the product $J\,\sigma y$. It should be noted that including effects of the background will alter this condition slightly.

We will hereforth assume y to be unity for both electron and photoabsorption experiments, but is should be noted that for γ radiation, y is the fluorescent yield which can be small for elements in the first two rows of the periodic table.

For spectral fine structure near an ionization edge, another quantity of interest is the sampling energy window since the signal to noise near an edge depends upon the energy resolution. Because a continuum of radiation is emitted from a synchrotron storage ring, the flux at the specimen depends upon the sampling window which is transmitted by the monochromator in front of the specimen. For electrons however, the sampling width need be limited only by the energy distribution of the electrons emitted from the source.

A factor which measures the sensitivity must include both the minimal amount of material detectable and the ability to detect spectral fine structure. We define the sensitivity to be the incident particle per unit area per unit time per unit energy bandwidth times the cross section for the process detected. For electrons, we define the sensitivity S_e to be

$$S_e = J_e \frac{d\,\sigma(E)}{dE} \tag{9}$$

where J_e is the incident current density in the beam (electrons/sec/unit area) and $d\,\sigma(E)/dE$ is the differential cross section per unit energy at a given loss E. In the same manner, for synchrotron radiation, we write

$$S_\gamma = J_\gamma \sigma_\gamma(E)/\Delta E \tag{10}$$

where $J_\gamma/\Delta E$ is the photon flux per ΔE bandwidth at the specimen and $\sigma_\gamma(E)$ is the photoabsorption cross section for incident photons of energy E. A useful figure of merit for comparing the two systems can be taken to be the ratio of the two sensitivity factors,

$$R = S_e/S_\gamma \tag{11}$$

B. Source Brightness

In order to determine the sensitivity factors, we must know the incident flux per unit energy bandwidth at the sample. In order to obtain this quantity the brightness of the two sources must be known (brightness/unit energy bandwidth). Brightness is an invariant in an optical system and gives us the additional information of how small an area of the sample we can probe.

The brightness for electron beam systems is only a property of the source and is independent of the energy lost by the electrons in traversing the sample. Presently there are three types of electron sources used in electron beam systems: the heated tungsten hairpin filament, the LaB_6 thermionic source and the field emission source. The thermionic tungsten source has a maximum brightness of 10^5A/cm^2/sterad,[11] while the LaB_6 source has a brightness one or two orders of magnitude higher.[12] Cold field emission sources can have a brightness greater[13] than 10^9A/cm^2/sterad. The energy spread for electrons leaving thermionic sources is 1-2 ev, while for cold field emission the energy spread can be as low as .25 ev.

In figure 2 we have plotted the brightness per unit energy bandwidth as a function of electron energy loss (in the case of electron sources) and as a function of photon energy (in the case of synchrotron sources). Unlike electron source brightness, photon beam brightness is dependent on the photon energy and drops rapidly for photon energies greater than a "cut-off" energy E_c, which is dependent upon certain parameters of the storage ring. Detailed discussions concerning the physics of producing "synchrotron light" can be found in the literature (eg ref. 2,3).

The main point to be noticed from fig.2 is that the brightness/unit energy bandwidth of proposed storage ring sources is several orders of magnitude lower than the brightness which can be achieved with electron sources at the present level of electron beam technology. In addition, since synchrotron radiation has a continuous energy distribution, it must pass through monochromators resulting in a loss of intensity. Electron beams can be fairly monochromatic as for example in field emitted electrons which have an energy distribution as small as 0.25 eV which corresponds to 0.001Å spread for photon wavelengths near the oxygen K-shell excitation edge.

The scattering cross sections for inner shell excitation and ionization by electrons is several orders of magnitude less than for photons, but the attainable flux at the sample compensates so that the sensitivity can be higher for electron beams than photon beams

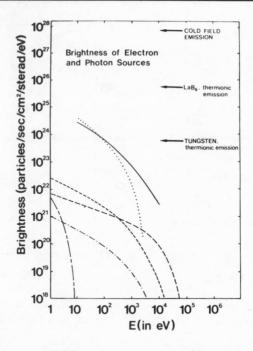

Fig. 2.

——— BNL National Light Source (X-ray) 2.5 GeV electrons, Effective source size $150\mu \times 6\mu$ (ref. 14) proposed.

····· BNL National Light Source (UV) 700 MeV electrons Source size $150\mu \times 6\mu$. (ref. 14) proposed.

— — DORIS 3.5 GeV electrons, .5A circulating current source size (effective) 1 mm $\times$ 10 mm (ref. 3,15).

----- SPEAR 3.0 GeV electrons, 35 A circulating current. Effective source size $415\mu \times 16.5\mu$ (ref. 16,17).

–·–·· SPEAR 3.0 GeV electrons, 200 mA circulating current. Effective source size 1 mm $\times$ mm (ref. 16, 17).

——— — The Sun (after ref. 15).

in some cases. Figures 3a, 3b show the ratio R of the sensitivities as a function of ionization (or excitation) energy. The horizontal axis is the energy lost by the incident electrons or the energy of the incident photons. R values greater than unity imply that the electron beam system has higher "sensitivity" than the photon beam system to which it is being compared.

C. Size of the Area Probed

In figure 3b, we considered an electron beam with a current density of $10^6 A/cm^2$ which corresponds to a 15Å diameter beam with a 20nA beam current.[18] Smaller diameter electron beams ($\sim$2.5Å) can be produced[19] (although with less current in the spot) so that it may be possible to use energy loss spectroscopy to examine local inhomogeneities at near atomic spatial resolution.

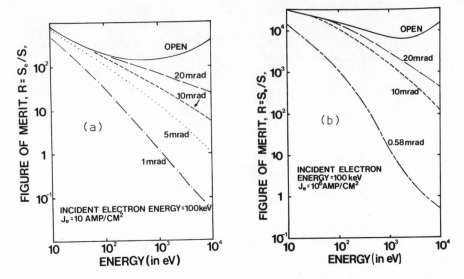

Fig. 3.   The relative detective sensitivities of electrons and photon beam systems.
$R = S_e/S_\gamma$ defined in eqn. 11. The comparison is for 100 keV incident electrons.

In fig. 3a we compare a conventional electron beam system with a current
density of 10 Amp/cm^2 at the sample with a practical photon beam system
(with monochromatization) from SPEAR which is used at the present. The
solid curve marked "open" corresponds to a comparison with an electron
beam system which collects all the electrons losing a particular energy, while
the dashed curves represent spectrometers which collect only electrons scattered through less than the angles indicated.

In fig. 3b we compare an electron beam system capable of producing
10^6Amp/cm^2 (see ref. 18) with a photon beam produced from the BNL
National Light Source (assuming a 2.5 GeV, 600 mA current, with a source
size $150\mu \times 6\mu$ and a beam divergence of 1 mrad $\times$ 0.1 mrad - see ref. 14).
We have assumed that all the radiation emanating from the storage ring
passes through a monochromater with 100% efficiency and is focused to a
spot size equal to the source size with no loss of intensity.

Producing small diameter photon beams from storage ring sources is more difficult
than the production of small diameter electron beams, because it is more difficult to
construct lenses for far UV and X-ray photons than it is for electrons. Since the radiation from the storage ring is a diverging beam, focusing is necessary to obtain a spot at
least as small as the apparent source size. It has been shown that photon beams of
sufficient intensity to obtain inner-shell spectra cannot be made much smaller[1] than
1000Å (this is because of the necessary monochromatization).

One advantage of using an electron beam system for obtaining spectral fine structure
(eg. EXAFS) is that energy loss measurements as a function of momentum transferred
can easily be achieved from small areas. In photon beam systems, one must perform

Compton scattering experiments in order to obtain information about momentum transfer as a function of excitation energy, and the probability of Compton scattering for photons less than 10keV is small compared to the probability of photoabsorption.[20]

In order to perform energy loss measurements as a function of momentum transfer with electron beams, a small spectrometer aperture must be used in order to obtain good momentum resolution. The momentum transfer depends upon the scattering angle as

$$\hbar q = P_0\sqrt{\theta^2 + \theta_E^2} \tag{12}$$

where P_0 is the incident electron momentum and $\theta_E = \dfrac{E}{P_0 V}$ where E is the energy loss and V the incident electron velocity. The precision with which the momentum transfer can be measured is $\Delta(\hbar q) = P_0 \Delta \theta$ or

$$\Delta q = \frac{2\pi}{\lambda_e}\Delta\theta \tag{13}$$

For 100 keV incident electrons, a spectrometer aperture subtending a half-angle of 1mrad at the specimen corresponds to a precision of $\Delta q = 0.17 \mathrm{\AA}^{-1}$ for a parallel incident beam.

If energy loss spectra are desired from selected microareas, the precision with which momentum transfer can be measured is coupled with the size of the area from which spectra are desired. For example, consider a diffraction limited system set to produce a small diameter beam. The spot size is given by

$$\delta = 0.61 \frac{\lambda_e}{\alpha} \tag{14}$$

where α is the semi-angle of the incident beam. For a spectrometer aperture equal to the illumination aperture (α), the minimum momentum precision (neglecting convolution effects) for a given spatial resolution (area illuminated) is given by

$$\Delta q = 0.61 \left[\frac{2\pi}{\delta} \right] \tag{15}$$

We see that if we require a momentum resolution of $0.1 \mathrm{\AA}^{-1}$ with 100 keV electrons, that we cannot do this from an area less than $38 \mathrm{\AA}$ diameter. It should be noted that when the spectrometer aperture is less than the illumination angle, the momentum resolution is restricted by the illumination angle.

A disadvantage of using a small aperture is that a large fraction of inelastically scattered electrons are not collected, effectively reducing the sensitivity factor. This effect has been included in figure 3 where it is seen that with even such a small aperture that electron beam systems are still competitive with photon beam systems.

In cases where spatial localization of spectroscopic information is not needed, then a small diameter probe is not necessary. For larger probe sizes, small concentrations of material can be analyzed since the total amount of material exceeds that given in Eq. 8. In this case, it is the total current onto the sample and not the current density which is important.

Existing storage ring facilities (eg. SPEAR) can produce fluxes in excess of 10^8 photons/sec in a 1.5 mm $\times$ 2 mm spot at the exit of the monochrometer with a 1 ev

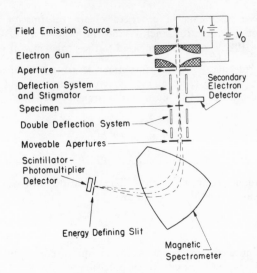

Fig. 4.   Schematic of a simple scanning transmission electron microscope. The two
            electrode gun accelerates and focuses the electron beam from the field emis-
            sion source to form a 100Å diameter probe at the specimen. By scanning
            the probe across the specimen with the upper deflection system, transmitted
            electrons can be used to modulate the intensity of a synchronously scanned
            display to form an image of the specimen. Transmitted electrons can be
            selected by the double deflection system, spectrometer and slit to obtain
            energy loss spectra and diffraction patterns. This microscope has been
            modified to produce a current density at the specimen of 10^6Amp/cm^2 (see
            ref. 18) by producing a 15Å diameter probe with a beam current of 20
            nAmps.

energy window. Values four to five orders of magnitude higher can be achieved in the
next generation storage ring sources. Because the cross sections for inner shell excitations are smaller for electrons, electron beam systems would require about four orders
of magnitude more probe current (100nA-10mA) in order to be competitive.

Such currents can be achieved in large spot sizes ($\sim$1 mm diameter) in electron
probe-type instruments and electron beam welding systems,[21] but it may be that the
Boersch effect is large making monochromatization necessary[22,23] (resulting in a reduction of intensity at the sample). In addition, with such large beam currents, specimen
heating may be a problem.

It appears that if spatial localization is not required that photon beam systems may
be superior for obtaining spectroscopic information. However, when spatial localization
is required, electron beam systems may be advantageous. It should be pointed out that
not only spectra can be obtained with electron beam systems, but also that imaging and
diffraction patterns can be easily obtained. As an example, fig. 4 shows the schematic
of an electron beam system[24] capable of imaging, obtaining spectra, and producing
diffraction patterns. The current density at the specimen in this system is $\sim10^2$
Amp/cm^2, with a probe size of $\sim$100Å. The advantage of having imaging capabilities

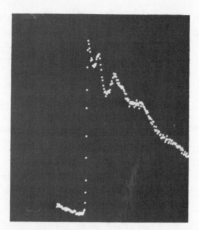

Fig. 5.   Spectrum of the carbon K edge from a $\sim$200Å thick graphite specimen. The spectrum was obtained in $<$ 10 min. with the microscope described in fig. 4. The full energy scale is 100 eV and the maximum peak height is 9.8 K. The beam energy was 17 keV.

is that spectra can be obtained from a given region, the specimen can be modified (e.g. intercalated), and the same area can then be reexamined. This has been done in the case of stage-one potassium intercalated graphite,[25] where the areas examined before intercalation were relocated to within a micron after intercalation. In addition, diffraction patterns were easily obtained in order to verify the structure of the sample.[25] Figure 5 shows the carbon K-edge of a 200Å-thick graphite specimen taken with this instrument.

IV. BEAM-INDUCED DAMAGE

It is important to compare the effect of damage by the beam on the ability to obtain spectral information before the incident radiation destroys or alters the sample being probed. Damage can arise either from bond breaking or distortion due to excitation, ionization or elastic nuclear collisions. Since the effects of elastic nuclear collisions are small for electrons with 100 keV energy or less, we will primarily be interested in beam damage from ionizing collisions.

For excitation energies less than 10 keV, the predominant interaction mechanism for photons is absorption via the photoelectric effect.[20] Electrons can either be elastically or inelastically scattered and generally only the inelastic collisions produce damage.

Consider the case where the cross section for damage is proportional to the K-shell cross section, such that each K-shell event has a certain probability, f, of being a damage event,

$$\sigma_{DAM} = f \, \sigma_K \tag{16}$$

If the source of the deposited energy is irrelevant, this fraction is the same for electrons and photons whose energy is above the excitation threshold. Thus, the same amount of damage will result per K-shell excitation for electrons and photons.

However, with electrons there are other excitations than the K-shell. The ratio between the probability for a K-shell excitation (or ionization) and any other excitation is

$$\frac{\sigma_K}{\sigma_{in}} \sim \frac{1}{Z^{5/2}} \tag{17}$$

Thus, there are many inelastic events for every K-shell event. If the cross-section for damage by electrons is related to the probability of having any energy loss event, f', then

$$\sigma_{DAM} = f'\sigma_K Z^{5/2} \tag{18}$$

so that for $f' > Z^{-5/2}$ there could be many damage events. Since there is an average energy loss per inelastic collision, this case represents damage being proportional to the total energy deposited.

In order to make an estimate of the effect, consider the ionization damage of carbon bonds. For electrons $(E_0 \geq 284\text{eV})$, $\sigma_K/\sigma_{in} \approx .01$. In many organic biological materials, the average energy lost by incident electrons per inelastic collision is about 37 eV, so that the number of K-shell excitations per eV lost by an incident electron is 3×10^{-4}. If it is assumed that the incident electron deposits all of the energy lost in the material, the G value (number of events per 100 eV deposited) is,

$$G = \frac{\sigma_K}{\sigma_{in}} \cdot \frac{10^2}{\bar{E}} = .03 \tag{19}$$

More sophisticated calculations[26] get a G value of $G = .025$ for 100 keV electrons incident on polyethylene and $G \approx 0.02$ for 100 keV electrons incident on various proteins.

A rough estimate of the corresponding G value for photoabsorption in the vicinity of the carbon K shell excitation edge can be obtained if it is assumed that the entire energy of the incident photon is deposited in the material. In this case.

$$G = \frac{100}{E} \tag{20}$$

Thus, at the K edge, G (284 eV) = .35, while for 500 eV photons G (500 eV) = .2. Therefore, photon beams will deposit almost one tenth less energy per K shell excitation than electron beams, because not every inelastic electron collision results in a K-shell excitation. One can show that the ratio of G values for particular excitations by electrons and photons is given approximately by,

$$\frac{G_n(e)}{G_n(\gamma)} \approx \left[\frac{E}{E_n}\right] \frac{5}{\bar{E}} \frac{Z_n}{\sqrt{Z}} \tag{21}$$

for electron energies between 30 keV - 200 keV and incident photon energies, E, not too distant from the excitation energy of interest (i.e. $E/E_n \sim 1-5$). Z_n is the number of electrons in the nth shell whose binding energy is E_n, $\bar{E}$ is the average energy lost by the incident electron and Z is the atomic number of the material. If we assume, for simplicity, that $\bar{E} \approx 5Z$, then

$$\frac{G_n(e)}{G_n(\gamma)} \approx \left[\frac{E}{E_n}\right] \frac{Z_n}{Z^{3/2}} \tag{22}$$

Since $Z_n \leq 2$ for the K-shell and $Z_n \leq 8$ for the L shell, for most elements of interest, fewer K and L shell excitations are produced per 100 eV deposited using electrons.

Thus we see that if the damage produced by ionizing radiation is proportional to the energy deposited and independent of the particular kind of incident radiation, then these G values give the relative amounts of damage production. For the case of carbon K-shell excitation, electrons are about ten times as damaging as photons whose energy is within a few hundred eV from the carbon K edge. However, if the damage is produced by a particular excitation, then not all inelastic electron scattering events will cause damage and electrons and photons will cause about the same amount of damage.

Therefore, in general, photons should produce less damage per inner shell excitation than electrons, provided that the photon energy is not much greater than the excitation energy of the edge being studied. In cases where the particular excitation being studied is responsible for the damage, photon and electron beams should be equally damaging.

V. OTHER COMPARISONS

A. Multiple Scattering

Multiple inelastic scattering is another area of concern when using electron beams for obtaining inner shell spectra. For electrons, the appropriate mean free path to consider for scattering events is that due to any inelastic event, while for photons, the mean free path for photoabsorption corresponds to the excitation explored. The electron mean free path is smaller than that for inner shell excitation so that there is an appreciable probability for an incident electron to lose energy in producing both an inner shell excitation and a valance shell excitation.

In cases where the probability for a double scattering event is comparable to single inner shell excitation events, the measured energy loss distribution does not represent the single scattering profile (which is proportional to ϵ_2 near the inner shell region), but is a convolution of the inner shell single scattering profile with lower lying loss events. There are several techniques for extracting the single scattering profile from measured electron energy loss spectra (e.g. ref. 4).

In order to minimize multiple scattering corrections it is necessary to use thinner samples in electron energy loss spectroscopy (EELS) experiments than with photoabsorption experiments. For EELS the optimum sample thickness for maximum signal to noise in an inner shell peak is about one mean free path for total inelastic scattering (if the sample is a small concentration in a matrix, it is the mean free path for the matrix material). Since, for photoabsorption experiments the optimum thickness is the mean free path for the excitation under study, the ratio of optimal thickness T_e and T_γ is

$$\frac{T_e}{T_\gamma} = \frac{510}{\ln\left[\dfrac{4 \times 10^5}{E}\right]} \frac{1}{Z^{5/2}} \tag{23}$$

for 100 keV electrons.

B. Parallel Data Collection

When recording spectral data with photon beams, the monochromator is scanned and the photoabsorption spectra of the sample is recorded in time sequential fashion.

Generally, electron energy loss spectra are also recorded in time sequential fashion by sweeping the energy loss spectrum across a slit. When an incident electron looses energy, producing an inner shell excitation, the energy lost can be anywhere from the inner shell edge to very large energies. Recording the spectrum serially would not record all the inner shell excitation events which passed through the spectrometer. For a given number of counts in the spectrum consisting of C channels, it is necessary to record the spectrum in a period C times longer than if the spectrum were recorded in parallel. This longer recording time implies that more energy is absorbed by the sample and is potential more damaging.

For EELS recorded in parallel, the ratio of total energy absorbed by the sample to obtain a given number of counts per channel (E_T^e) to that for photons (E_T^γ) is nearly inversely proportional to the ratio of the respective G values,

$$\frac{E_T^e}{E_T^\gamma} \approx \frac{G_\gamma}{G_e} \tag{24}$$

If the spectrum is recorded serially, the ratio is increased by at least C.

VI. CONCLUSION

Clearly, the foregoing comparison is not complete. We have not discussed polarization effects or the ability to perform experiments involving transient phenomena or analysis of objects in situ. Other comparisons we have avoided are the relative sizes of the two types of instruments and their comparative cost.

From the comparisons we have made, prudence should be employed in interpreting the results beyond what has been intended; the general nature of the comparison is only to illustrate trends.

the trends we find are:

1. for interest in spectroscopic information from small amounts of material in micro-areas, electron beam systems should have a higher detection sensitivity than photon beam systems, but

2. for interest in information in a statistically averaged sense, photon beams may be better;

3. electron beam systems have the ability to obtain easily spectra as a function of momentum transfer but

4. because other excitations besides the one of interest occur, EELS may be more damaging to the sample. This depends upon the exact mechanism of damage, and in some cases the damage may be equal for the two systems.

ACKNOWLEDGEMENTS

This work was supported by the U.S. Dept. of Energy and the National Science Foundation.

REFERENCES

1. This chapter is almost entirely drawn from the material in the paper by M. Isaacson and M. Utlaut, Optik **50,** 213 (1978).

2. M. L. Perlman, E. M. Rowe and R. E. Watson, Physics Today, July 1974, p. 30.

3. E. M. Rowe and J. H. Weaver, Sci. Amer. **236,** 32 (1977).

4. J. Daniels, C. V. Festenberg, H. Raether and K. Zeppenfeld in Springer Tracts in Modern Physics **54** (1970) p. 78, ed. G. Hohler.

5. For an excellent overview of electron energy loss spectroscopy see D. C. Joy and D. M. Maher, Science **206,** 162 (1979).

6. M. Born and E. Wolf, Principles of Optics, Pergamon Press, Oxford (1959) p. 610.

7. J. Hubbard, Proc. Phy. Soc. (London) **A68,** 976 (1955).

8. D. Pines, Elementary Excitations in Solids, W. A. Benjamin, San Francisco (1974), Ch. 4.

9. N. Swanson and C. Powell, Phys. Rev. **167,** 592 (1968).

10. C. Powell, Rev. Mod. Phys. **48,** 33 (1976).

11. M. E. Haine and P. A. Einstein, Brit. J. Appl. Phys. **3,** 40 (1952).

12. A. N. Broers in Fifth International Conference on Ion and Laser Beam Science and Technology, (1972) ed. R. Bakish.

13. A. V. Crewe, Prog. in Optics **11** (1973) 225. (ed. E. Wolf), North Holland, Amsterdam.

14. J. P. Belwett, ed. Proposal for a National Synchrotron Light Source. Brookhaven National Laboratory Report BNL 50595 (1977).

15. C. Kunz in Vacuum Ultraviolet Radiation Physics p. 753 (1975), Pergamon-Vieweg, Braunschweig (eds. E. E. Koch, R. Haensel and K. Kunz). S. Doniach, I. Lindau, W. E. Spicer and H. Winick, J. Vac. Sci. Tech. **12,** 1123 (1975).

16. H. Winick in Prox. IX International Accelerator Conference, Stanford University, SLAC 2-7 May 1974, p. 685.

17. S. Doniach, I. Lindau, W. E. Spicer and H. Winick, J. Vac. Sci. Tech. **12,** 1123 (1975).

18. A design for such an instrument can be found in N. W. Parker, M. Utlaut, M. Isaacson and A. V. Crewe, Proc. 37th Annual EMSA Meeting, San Antonio, ed G. W. Bailey (Claitors Press: Baton Rouge) 1979.

19. A. V. Crewe and J. Wall, Optik **30,** 461 (1970).

20. W. Heitler, The Quantum Theory of Radiation, 3rd Ed. (Oxford at the Clarendom Press, Oxford: 1954) Ch. 21-22.

21. P. Grivet, Electron Optics, 2nd Edition (1972) (Pergamon Press Oxford) Ch. 22.

22. H. Boersch, Z. Physik **139,** 115 (1954).

23. A. V. Crewe, Optik **50,** 205 (1978).

24. A. V. Crewe, M. Isaacson, D. Johnson, Rev. Sci. Inst. **40,** 241, (1969).

25. D. M. Hwang, M. Utlaut, M. Isaacson, S. Solin, Phys. Rev. Lett. **43,** 882 (1979).

26. J. Durup and R. L. Platzman, Int. J. Rad. Phys. Chem. **7,** 121 (1975).

SOME THOUGHTS CONCERNING THE RADIATION DAMAGE RESULTING FROM MEASUREMENTS OF INNER SHELL EXCITATION SPECTRA USING ELECTRON AND PHOTON BEAMS*

M. S. Isaacson

School of Applied and Engineering Physics
Cornell University
Ithaca, New York   14853

In an earlier paper,[1] we compared the efficiency of electron and photon beams for determining microchemical environment. The conclusion, that electron beam systems might have higher sensitivity than a photon beam system when analyzing submicron size areas, is in fair agreement with the assessment by J. Kirz[2] in which various beam spectroscopies were compared with regard to elemental identification of biomaterials. Estimates concerning the relative damage imparted to the sample by ionizing beams were made in (1) and (2) using somewhat different assumptions. However, no effort was made to assess the radiation damage due to a spectroscopic measurement. [Although in (3), comparisons (for electrons) of doses necessary to acquire given statistics in a spectroscopic measurement were compared to the doses at which manifestations of damage occurred.]

It is the purpose of this short note to try to compare the relative amount of beam induced damage that is imparted to a sample in the course of obtaining an inner shell excitation spectra using either a photon or electron beam as the excitation source. The situation is quite complex in that there are various types of damage, the damage is sample dependent and conditions for obtaining optimum signal to noise in the spectra need not be the same when using electron or photon beams. Therefore, I should stress at the outset that I am only attempting to make an order of magnitude estimate of the relative induced damage.

To simplify matters, I will limit the discussion to two types of samples: 1) an organic sample in which we are interested in EXAFS information concerning the carbon, nitrogen or oxygen atoms (i.e., using photon energies or electron energy losses of several hundred eV) and: 2) an organometallic sample in which we are interested in the metal atom environment (i.e., we want to obtain spectra near the inner shell ionization edge of the metal atom). For our purposes, we consider iron in a carbon-like matrix

---

* Supported by the NSF through the Cornell University, Materials Science Center.

and will evaluate the damage due to measurements in the vicinity of the L_{23} edge ($\sim$700 eV) and the K edge ($\sim$7000 eV). We want to evaluate the number of information containing excitations that one obtains to the number of excitations which result in damage to the sample (see (1)).

There is little intrinsic difference in the nature of the excitations induced by fast electrons from those induced by photons in the 15-30 eV energy range since the bulk of the oscillator strength for valence shell excitation lies in that energy range (eg. 4,5). The end result of damage is basically the same if the sample is irradiated by 30 eV photons or 30 KeV electrons. In fact, irradiation by higher energy photons (100 eV to 10 KeV) can be thought of primarily as a means of producing the lower energy electrons (secondaries, Auger, etc.) which cause damage. For example, in polyethylene, the G values for damage (the number of damage events per 100 eV absorbed) is essentially the same for 10 keV photons as for 100 KeV electrons.[6]

The quantity of interest in this paper is not the type of radiation damage, but rather the number of information producing excitations which one obtains for every excitation which results in damage. A prime quantity to calculate is the ionization damage of carbon bonds in an organic material. We will use as our indicator of damage the number of carbon K shell excitations. This is not unreasonable since such an excitation certainly results in some form of damage of organic materials via ionization on the resulting secondary electrons.[6] We therefore define our G value for damage to be the number of carbon K shell excitations produced for every 100 eV deposited in the sample G_{ck}.

For electrons, this becomes

$$G_{ck}(e^-) = \frac{\sigma_k}{\sigma_{in}} \frac{100}{\bar{E}} \tag{1}$$

where σ_k is the carbon K shell excitation cross-section (for electrons). σ_{in} is the total cross-section for inelastic electron scattering and $\bar{E}$ is the average energy lost per inelastic collision (in eV). For 100 KeV electrons incident on carbon, $\sigma_k/\sigma_{in} = .025$ whereas for various organic materials $\sigma_k/\sigma_{in} = .01$.[7] For most biological materials, $\bar{E} = 37$ eV,[7] so that $G_{ck}(e^-) \approx .05$ for the samples we are concerned about in this paper. More exact calculations in ref. 6 that take into account the distribution of primary energy losses and the distribution of the secondary products give $G_{ck}(e^-) = .02$ for 100 KeV electrons incident on various proteins and polyethylene. In that same paper, a calculation of the G value for carbon K shell excitation by high energy photons incident on polyethylene gives $G_{ck}(\gamma) = .03$ for 10 KeV photons, thus indicating comparable damage induced by the electrons or photons.

We are concerned with photon energies near 300 eV, and around 700 eV (near the iron L_{23} edge) and 7000 eV (near the iron K edge). Extrapolating the G values from reference 6 gives $G_{ck}(\gamma) = .031, .040$ and $.045$ for photons of energy near the Fe K edge, the Fe L_{23} edge and the carbon K edge respectively. However, extrapolation to the several hundred eV range may not be very accurate. We can get another estimate of $G_{ck}(\gamma)$ near the carbon K edge assuming all the photon energy to be deposited in the sample. Then, $G_{ck}(\gamma) = 100/E_\gamma$.[1] This would give a slightly higher G_{ck} value than the extrapolated value. Given the difference in the two results, it is clear that there is probably no substantial difference between the G values for damage by 100 KeV electrons and photons of several hundred eV energy.

Let us now consider the damage induced by obtaining spectra due to iron atoms in a carbon-like matrix. Here we want to evaluate the relative excitation cross-sections for iron K and L_{23} excitations compared to carbon K shell excitation. For 100 KeV electrons, this ratio $\sigma_{L_{23}}(Fe)/\sigma_k(C) \approx \frac{1}{4}$ as determined by experiment[8] or using Bethe-type cross-section expressions.[9,10] Therefore, one gets more damage when obtaining spectra near the iron L_{23} edge than when obtaining spectra near the carbon (or nitrogen or oxygen) K edges. Since the photo absorption cross-section is essentially proportional to the photon energy times the electron differential cross section,[9] then $\sigma_{L_{23}}(Fe)/\sigma_K(C)$ will be slightly larger for photons than for the 100 keV electrons.[11] This means less of an increase in damage when obtaining iron L_{23} spectra as opposed to carbon K spectra. However, for our purposes here there is no significant difference.

It is easier to measure EXAFS spectra near the iron K edge rather than near the L edges when using photons. (This is opposite of the case with electrons due to the fact that the cross-section for the Fe K excitations near 7 KeV is very much smaller than for the L_{23} excitation when using 100 KeV electrons). Since the K shell cross-section is less than the L shell cross section for the case of iron when using either photons or electrons, it is clear that more damage will be induced when using 7 KeV photons than 700 eV photons. From ref. 6 we would get $G_{ck}(\gamma) = 0.031$ for 7 KeV photons incident on polyethylene. This is not significantly different than the G_{ck} for 100 KeV electrons given earlier.

So far, we have not explicitly taken into account the spectral background (i.e. the background preceding an ionization peak). The following arguments should give us a very rough estimate of this effect on damage. For maximum signal to noise in the spectra, the optimal thickness for obtaining carbon K shell excitations spectra with 100 KeV electrons is about 3/4 the optimal thickness using photon beams (around several hundred eV).[1,11] Neglecting the sharp excitonic-like structure within 10 eV of the excitation edge, the edge jump ratio is about the same in the two cases [about $10^{(12)}$] When the sample gets thicker than this optimum value, multiple scattering becomes a factor in the background when using electron beams. For example, samples that are twice as thick as the optimum thickness might have jump ratios as low as two to one or less. Thus, to obtain the same statistical information one might need two to five times the total number of counts thereby increasing the damage by that amount. In addition, the probability of non-information containing events per carbon K shell excitation also increases, thus further increasing the damage. Thus, when using samples much thicker than the optimum thickness, one could get more damage using electron beams than with photon beams when trying to obtain an inner shell spectra. A more accurate assessment of the role of the background in increasing the induced beam damage is a subject of further investigations.

Therefore, within the context of our assumptions and approximations, it would appear that within a factor of two or so, there is no significant difference in the damage incurred by the sample due to the measurement of inner shell spectra of organic samples when using photons or 100 KeV electron beams. It would appear that to get a more accurate assessment comparative measurements on similar samples would be appropriate.

Acknowledgements: This work was supported by the NSF through the Cornell University Materials Science Center. I would like to thank D. Joy for the opportunity to reflect upon the subject.

References

1. M. Isaacson and M. Utlaut. Optik **50,** 213 (1978).

2. J. Kirz, "Specimen Damage Considerations in Biological Microprobe Analysis" in SEM/80. (SEM, Inc. AMF O'hare). O. Johari, Ed. (1980) in press.

3. M. Isaacson, Ultramicroscopy **4,** 193 (1979).

4. F. Williams, "Early Processes in Radiation Chemistry and Reactions of Intermediates" in *Radiation Chemistry of Macromolecules* I. (Academic Press, New York and London). M. Dole, ed. (1972). p. 7-23.

5. R. L. Platzman, in: *Radiation Research.* (North Holland, Amsterdam) G. Silini, ed. (1967). p. 20-42.

6. J. Durup and R. L. Platzman, Int. J. Rad. Phy. Chem. **7,** 121 (1975).

7. M. Isaacson, D. Johnson and A. V. Crewe, Rad. Res. **55,** 205 (1973).

8. P. Trebbia, These Universite Paris-Sud, Centre d'Orsay. Nov. 26, 1979.

9. M. Inokuti, Rev. Mod. Phys. **43,** 297 (1971).

10. C. J. Powell, Rev. Mod. Phys. **48,** 33 (1976).

11. M. Isaacson and D. Johnson, Ultramicroscopy, **1,** 33 (1975).

12. J. H. Hubbell, Atomic Data, **3,** 241 (1971).

INDEX